面向新工科普通高等教育系列教材

人工智能英语教程

张强华　卓　勤　司爱侠　编著

机械工业出版社

本书内容包括人工智能概述、人工智能的利与弊、知识表示和知识图谱、人工智能中的推理、模糊逻辑系统、知情搜索算法和不知情搜索算法、专家系统、计算机视觉、机器学习和深度学习、十大机器学习算法、人工神经网络、用于大数据的机器学习、模式识别、自然语言处理、人工智能中的智能体、智能制造、用于人工智能的五种编程语言、人工智能平台、人工智能的十大趋势、人工智能在云计算中的作用等。

全书共 12 个单元，每一单元包括课文、单词、词组、缩略语、参考译文、习题。附录 A、附录 B、附录 C 既可用于复习和背诵，又可作为小词典查阅。本书课文均配有音频材料，扫描二维码即可收听。

本书既可作为高等院校本、专科人工智能及相关专业的专业英语教材，也可作为相关从业人员自学的参考书，还可作为培训班的培训用书。

本书配有授课电子课件，需要的教师可登录 www.cmpedu.com 免费注册，审核通过后下载，或联系编辑索取（微信：15910938545，电话：010-88379739）。

图书在版编目（CIP）数据

人工智能英语教程 / 张强华，卓勤，司爱侠编著. —北京：机械工业出版社，2022.8

面向新工科普通高等教育系列教材

ISBN 978-7-111-71159-9

Ⅰ. ①人… Ⅱ. ①张… ②卓… ③司… Ⅲ. ①人工智能-英语-高等学校-教材 Ⅳ. ①TP18

中国版本图书馆 CIP 数据核字（2022）第 115094 号

机械工业出版社（北京市百万庄大街 22 号 邮政编码 100037）
策划编辑：郝建伟 责任编辑：郝建伟 解 芳
责任校对：张艳霞 责任印制：邸 敏

北京富资园科技发展有限公司印刷

2022 年 8 月第 1 版·第 1 次印刷

184mm×260mm·16.25 印张·399 千字

标准书号：ISBN 978-7-111-71159-9

定价：69.00 元

电话服务　　　　　　　　　　　网络服务

客服电话：010-88361066　　　机 工 官 网：www.cmpbook.com
　　　　　010-88379833　　　机 工 官 博：weibo.com/cmp1952
　　　　　010-68326294　　　金 书 网：www.golden-book.com
封底无防伪标均为盗版　　　机工教育服务网：www.cmpedu.com

前　言

2017 年，国务院印发了《新一代人工智能发展规划》，提出了六方面的重点任务和一系列保障措施，要求到 2020 年，我国人工智能核心产业规模超过 1500 亿元人民币，带动相关产业规模达 1 万亿元；到 2025 年，人工智能核心产业规模超过 4000 亿元，带动相关产业规模达 5 万亿元；到 2030 年，产业规模超过 1 万亿元，带动相关产业规模达 10 万亿元，我国人工智能理论、技术与应用总体达到世界领先水平。我国人工智能技术已经进入了高速发展阶段，各行各业对相关人才的需求十分旺盛。许多高校陆续开设了人工智能及相关专业，培养社会急需的专业人才。

人工智能技术发展迅猛的特点，要求从业人员必须掌握许多新技术、新方法，由此也提升了对其专业英语水平的要求。在职场中，具备相关职业技能并精通专业外语的人员往往能赢得竞争优势，成为不可或缺的甚至是引领性的人才。本书旨在帮助读者提高人工智能专业英语水平。

本书具有如下特色。

1）选材广泛，内容包括人工智能概述、人工智能的利与弊、知识表示和知识图谱、人工智能中的推理、模糊逻辑系统、知情搜索算法和不知情搜索算法、专家系统、计算机视觉、机器学习和深度学习、十大机器学习算法、人工神经网络、用于大数据的机器学习、模式识别、自然语言处理、人工智能中的智能体、智能制造、用于人工智能的五种编程语言、人工智能平台、人工智能的十大趋势、人工智能在云计算中的作用等。

2）体例创新，非常适合教学。本书共 12 个单元，每一单元包含以下部分。课文——选材广泛、风格多样、切合实际的两篇专业文章；单词——给出课文中出现的新词，读者由此可以积累人工智能专业的基础词汇；词组——给出课文中的常用词组；缩略语——给出课文中出现的、业内人士必须掌握的缩略语；参考译文——有助于读者对照理解课文，并提高读者的翻译能力；习题——包括针对课文的练习、词汇练习、翻译练习。附录 A、附录 B、附录 C 既可用于复习与背诵，又可作为小词典长期查阅。

3）习题量适当，题型丰富，难易搭配，便于教师组织教学。

4）教学资源丰富，本书提供配套的教学大纲、教学 PPT、参考答案及参考试卷等资料，需要的教师可登录 www.cmpedu.com 免费注册，审核通过后下载。

5）本书课文均配有音频材料，扫描每单元的二维码下载后即可收听（音频建议用耳机收听）。

6）编者有近 20 年 IT 行业英语图书的编写经验。在编者编写的书籍中，有三部"十一五"国家级规划教材，一部全国畅销书，一部华东地区教材二等奖图书。这些编写经验有助于本书的完善与提升。

本书使用的是英式音标。

在使用本书的过程中，读者可以通过电子邮件与编者交流。邮件标题请注明姓名及"索取机械工业出版社人工智能英语参考资料"字样。编者邮箱为 zqh3882355@sina.com 和 zqh3882355@163.com。

编　者

目　　录

Unit 1

Text A

扫码听课文

Overview of Artificial Intelligence

1. What is artificial intelligence

The answer to this question would depend on who you ask. A layman, with a fleeting understanding of this technology, would link it to robots. If you ask about artificial intelligence to an AI researcher, he would say that it's a set of algorithms that can produce results without having to be explicitly instructed to do so. Both of these answers are right. So to summarize, the definition of artificial intelligence is as follows.

- An intelligent entity created by humans.
- Capable of performing tasks intelligently without being explicitly instructed.
- Capable of thinking and acting rationally and humanely.

At the core of artificial intelligence, it is a branch of computer science that aims to create or replicate human intelligence in machines. But what makes a machine intelligent? Many AI systems are powered with the help of machine learning and deep learning algorithms. AI is constantly evolving, what was considered to be part of AI in the past may now just be looked at as a computer function. For example, a calculator may have been considered to be a part of AI in the past. Now, it is considered to be a simple function. Similarly, there are various levels of AI.

2. Why is artificial intelligence important

The goal of artificial intelligence is to aid human capabilities and help us make advanced decisions with far-reaching consequences. From a technical standpoint, that is the main goal of AI. When we look at the importance of AI from a more philosophical perspective, we can say that it has the potential to help humans live more meaningful lives that are devoid of hard labour. AI can also help manage the complex web of interconnected individuals, companies, states and nations to function in a manner that's beneficial to all of humanity.

Currently, artificial intelligence is shared by all the different tools and techniques that have been invented by us over the last thousand years to simplify human effort, and to help us make better decisions. Artificial intelligence is one such creation that will help us in further inventing ground-breaking tools and services that would exponentially change the way we live, by hopefully removing strife, inequality and human suffering.

We are still a long way away from those kinds of outcomes. But it may come around in the future. Artificial intelligence is currently being used mostly by companies to improve their process efficiencies, automate resource-heavy tasks and make business predictions based on data available to us. As you see, AI is significant to us in many ways. It is creating new opportunities in the world, helping us improve our productivity, and so much more.

3. Levels of artificial intelligence

(1) Artificial Narrow Intelligence (ANI)

Also known as narrow AI or weak AI, artificial narrow intelligence is goal-oriented and is designed to perform singular tasks. Although these machines are seen to be intelligent, they function under minimal limitations, and thus, are referred to as weak AI. It does not mimic human intelligence; it stimulates human behaviour based on certain parameters. Narrow AI makes use of NLP (Natural Language Processing) to perform tasks. This is evident in technologies such as chatbots and speech recognition systems such as Siri. Making use of deep learning allows you to personalise user experience, such as virtual assistants who store your data to make your future experience better.

Examples of weak or narrow AI: Siri, Alexa, Cortana, IBMs Watson, self-driving cars, facial recognition softwares, E-mail spam filters and prediction tools.

(2) Artificial General Intelligence (AGI)

Also known as strong AI or deep AI, artificial general intelligence refers to the concept through which machines can mimic human intelligence while showcasing the ability to apply their intelligence to solve problems. Scientists have not been able to achieve this level of intelligence yet. Significant research needs to be done before this level of intelligence can be achieved. Scientists would have to find a way through which machines can become conscious through programming a set of cognitive abilities. A few properties of deep AI are as follows.

- Recognition.
- Recall.
- Hypothesis testing.
- Imagination.
- Analogy.
- Implication.

It is difficult to predict whether strong AI will continue to advance or not in the foreseeable future, but with speech and facial recognition continuously showing advancements, there is a slight possibility that we can expect growth in this level of AI too.

(3) Artificial Super-Intelligence (ASI)

Currently, Super-Intelligence is just a hypothetical concept. People assume that it may be possible to develop such an artificial intelligence in the future, but it doesn't exist in the current world. Super-Intelligence can be known as that level wherein the machine surpasses human capabilities and becomes self-aware. This concept has been the muse to several films and science fiction novels wherein robots who are capable of developing their feelings and emotions can overrun humanity itself. It would be able to build emotions of its own, and hypothetically, be better than humans at art, sports, math, science and more. The decision-making ability of a Super-Intelligence would be greater than that of a human being. The concept of Artificial Super-Intelligence is still unknown to us, its consequences can't be guessed, and its impact cannot be measured just yet.

4. Goals of artificial intelligence

So far, you've seen what AI means, the different levels of AI and its applications. But what are the goals of AI? What is the result that we aim to achieve through AI? The overall goal would be to allow machines and computers to learn and function intelligently. Some of the other goals of AI are as follows.

(1) Problem-solving

Researchers developed algorithms that were able to imitate the step-by-step process that humans use while solving a puzzle. In the late 1980s and 1990s, research had reached a stage wherein methods had been developed to deal with incomplete or uncertain information. But for difficult problems, there is a need for enormous computational resources and memory power. Thus, the search for efficient problem-solving algorithms is one of the goals of artificial intelligence.

(2) Knowledge representation

Machines are expected to solve problems that require extensive knowledge. Thus, knowledge representation is central to AI. AI represents objects, properties, events, cause, effect and much more.

(3) Planning

One of the goals of AI should be to set intelligent goals and achieve them. AI is able to make predictions about how actions will impact change, and what are the choices available. An AI agent will need to assess its environment and accordingly make predictions. This is why planning is important and can be considered as a goal of AI.

(4) Learning

One of the fundamental concepts of AI, machine learning, is the study of computer algorithms that continue to improve over time through experience. There are different types of ML. The commonly known types of are unsupervised machine learning and supervised machine learning.

(5) Social intelligence

Affective computing is essentially the study of systems that can interpret, recognize and process human behavior. It is a confluence of computer science, psychology and cognitive science. Social intelligence is another goal of AI as it is important to understand these fields before building algorithms.

Thus, the overall goal of AI is to create technologies that can incorporate the above goals and create an intelligent machine that can help us work efficiently, make decisions faster and improve security.

5. Future of artificial intelligence

We have always been fascinated by technological changes. Currently, we are living amidst the greatest AI advancements in our history. Artificial intelligence has emerged to be the net greatest advancement in the field of technology. This has not only impacted the future of every industry, but has also acted as the driver of emerging technologies such as big data, robotics and IoT. At that rate at which AI is advancing, there is no doubt that it will continue to flourish in the future. With the advancement of

AI and its technologies, there will be a greater need for skilled professionals in this area.

An AI certification will give you an edge over other participants in the industry. As facial recognition, AI in healthcare, chat-bots continue to grow, now it would be the right time to work towards building a successful AI career. Virtual assistants are already part of our everyday life without us knowing it. Self-driving cars by tech giants like Tesla have shown us a glimpse of what the future will look like. There are so many more advancements to be discovered, and this is only the beginning. According to the World Economic Forum, 133 million new jobs are said to be created by artificial intelligence by the year 2024. The future of AI is definitely bright.

✎ New Words

intelligence	[ɪn'telɪdʒəns]	n.智能，智力
layman	['leɪmən]	n.门外汉，外行
researcher	[rɪ'sɜ:tʃə]	n.研究者
instruct	[ɪn'strʌkt]	vt.教导，指导
entity	['entəti]	n.实体
task	[tɑ:sk]	n.工作，任务 vt.交给某人（任务）
function	['fʌŋkʃn]	n.功能；函数
calculator	['kælkjuleɪtə]	n.计算器；计算者
aid	[eɪd]	n.&v.帮助，辅助
capability	[ˌkeɪpə'bɪləti]	n.能力
decision	[dɪ'sɪʒn]	n.决策，决定
standpoint	['stændpɔɪnt]	n.立场，观点
perspective	[pə'spektɪv]	n.观点，看法
meaningful	['mi:nɪŋfl]	adj.意味深长的；有意义的
web	[web]	n.网络
interconnect	[ˌɪntəkə'nekt]	vi.互相连接，互相联系 vt.使互相连接；使互相联系
beneficial	[ˌbenɪ'fɪʃl]	adj.有益的，有帮助的
invent	[ɪn'vent]	vt.发明，创造
ground-breaking	['graʊnd breɪkɪŋ]	adj.开拓性的，独创的
prediction	[prɪ'dɪkʃn]	n.预测，预报
available	[ə'veɪləbl]	adj.可用的，可获得的；有空的
singular	['sɪŋgjələ]	adj.单个的
limitation	[ˌlɪmɪ'teɪʃn]	n.限制，局限；极限
mimic	['mɪmɪk]	vt.模仿
stimulate	['stɪmjuleɪt]	vt.刺激；激励；促进

parameter	[pə'ræmɪtə]	n.参数，参量；限制因素，决定因素
recognition	[ˌrekəg'nɪʃn]	n.识别
personalise	['pɜːsənəlaɪz]	vt.个性化；个人化
virtual	['vɜːtʃuəl]	adj.（计算机）虚拟的；实质上的，事实上的
showcase	['ʃəʊkeɪs]	v.展示（优点） n.（商店或博物馆等的）玻璃柜台，玻璃陈列柜；展示（优点的）场合
ability	[ə'bɪləti]	n.能力；才能
apply	[ə'plaɪ]	v.应用，适用
solve	[sɒlv]	v.解决；破解
property	['prɒpəti]	n.特性，属性
imagination	[ɪˌmædʒɪ'neɪʃn]	n.想象；想象力
analogy	[ə'nælədʒi]	n.类推
implication	[ˌɪmplɪ'keɪʃn]	n.含义，蕴涵，蕴含
continue	[kən'tɪnjuː]	vi.持续，连续
foreseeable	[fɔː'siːəbl]	adj.可预见到的
possibility	[ˌpɒsə'bɪləti]	n.可能性；机会；潜力
hypothetical	[ˌhaɪpə'θetɪkl]	adj.假设的，假定的，假想的
surpass	[sə'pɑːs]	vt.超过，优于，胜过
self-aware	[ˌself ə'weə]	adj.自知的，自我意识的
consequence	['kɒnsɪkwəns]	n.结果；重要性
imitate	['ɪmɪteɪt]	vt.模仿，效仿
puzzle	['pʌzl]	n.谜；疑问；智力游戏 v.迷惑；苦苦思索
stage	[steɪdʒ]	n.阶段
enormous	[ɪ'nɔːməs]	adj.巨大的；极大的
knowledge	['nɒlɪdʒ]	n.知识
fundamental	[ˌfʌndə'mentl]	adj.基础的
affective	[ə'fektɪv]	adj.情感的
confluence	['kɒnfluəns]	n.（事物的）汇合；汇流
psychology	[saɪ'kɒlədʒi]	n.心理学
fascinate	['fæsɪneɪt]	v.深深吸引；迷住
advancement	[əd'vɑːnsmənt]	n.发展，推动
emerge	[i'mɜːdʒ]	v.出现，兴起
flourish	['flʌrɪʃ]	vi.繁荣；活跃
glimpse	[glɪmps]	n.一瞥，扫视

✎ Phrases

depend on	根据；依据，依靠
artificial intelligence	人工智能（AI）
a set of	一套，一组
computer science	计算机科学
machine intelligent	机器智能
machine learning	机器学习（ML）
deep learning	深度学习
be based on	基于
weak AI	弱人工智能
be referred to as	被称作，被称为
speech recognition	语音识别
user experience	用户体验
virtual assistant	虚拟助手
facial recognition	面部识别
strong AI	强人工智能
deep AI	深度人工智能
hypothesis testing	假设检验
science fiction novel	科幻小说
computational resource	计算资源
memory power	存储能力
knowledge representation	知识表示
unsupervised machine learning	无监督机器学习
supervised machine learning	监督机器学习
social intelligence	社会智能
affective computing	情感计算
big data	大数据
self-driving car	自动驾驶汽车

✎ Abbreviations

ANI (Artificial Narrow Intelligence)	窄人工智能
NLP (Natural Language Processing)	自然语言处理
AGI (Artificial General Intelligence)	通用人工智能
ASI (Artificial Super-Intelligence)	超人工智能
IoT (Internet of Things)	物联网

Reference Translation

人工智能概述

1. 什么是人工智能

这个问题的答案取决于你问谁。一个外行人一想到这个技术，可能会将其与机器人联系起来。如果你向人工智能研究人员询问人工智能，他会说这是一组算法，可以产生结果而无须明确指示要这样做。这两个答案都是正确的。综上所述，人工智能的定义如下。

- 由人类创建的智能实体。
- 能够在没有明确指示的情况下智能地执行任务。
- 能够理性并像人类一样思考和行动。

人工智能的核心是计算机科学的一个分支，目的是在机器中创建或复制人类智能。但是，是什么使机器智能化？许多人工智能系统都借助机器学习和深度学习算法来提供支持。人工智能在不断发展，过去被视为人工智能的一部分，现在可以看作是一种计算机功能。例如，过去可能认为计算器是人工智能的一部分；现在，它被认为是一个简单的功能。同样，有各种级别的人工智能。

2. 为什么人工智能很重要

人工智能的目标是提高人的能力，并帮助我们做出具有深远影响的高级决策。从技术角度来看，这是人工智能的主要目标。当从哲学的角度看待人工智能的重要性时，我们可以说它有潜力帮助人们过上更有意义的生活，而无须辛苦劳动。人工智能还可以帮助管理相互联系的个人、公司、州和国家（或地区）组成的复杂网络，从而以对全人类有利的方式发挥作用。

当前，人工智能已被我们过去一千年来发明的所有不同工具和技术所共享，这些工具和技术可简化人类的工作并帮助我们做出更好的决策。人工智能就是这样一种创造，它将帮助我们进一步发明开创性的工具和服务，有望通过消除冲突、不平等和人类苦难，以指数方式改变我们的生活方式。

我们离这些结果还有很长的路要走。但是将来可能会出现。目前，各公司主要将人工智能用于提高流程效率、自动执行资源丰富的任务，以及根据我们可用的数据进行业务预测。如你所见，人工智能在许多方面对我们都很重要。它在世界上创造了新的机会，帮助我们提高了生产力，如此等等。

3. 人工智能等级

（1）窄人工智能（ANI）

窄人工智能也称为狭义人工智能或弱人工智能，它是面向目标的，旨在执行单个任务。尽管这些机器被认为是智能的，但它们在最小的局限下运行，因此被称为弱人工智能。它不模仿人类的智力，而根据某些参数刺激人类行为。窄人工智能利用 NLP（自然语言处理）来执行任务。这在诸如聊天机器人之类的技术和诸如 Siri 之类的语音识别系统中很明显。利用深度学习，你可以个性化用户体验，例如用虚拟助手来存储数据以改善你的未来体验。

弱人工智能的示例有：Siri、Alexa、Cortana、IBM 的 Watson、自动驾驶汽车、面部识别软件、垃圾电子邮件过滤器和预测工具。

（2）通用人工智能（AGI）

通用人工智能也被称为强人工智能或深度人工智能，这个概念是指机器可以模仿人类智能，同时展示应用其智能解决问题的能力。科学家们尚未能够实现这种等级的智能。在达到这种智能水平之前，需要进行大量研究。科学家将必须找到一种方法：通过编程设定一组认知能力，让机器能够变得有意识。深度人工智能的一些特性如下。

- 认出。
- 记起。
- 假设检验。
- 想象力。
- 类推。
- 蕴含。

很难预测强人工智能在可预见的未来是否会继续发展，但是随着语音和面部识别技术的不断发展，也许我们可以期待此类人工智能的水平会增长。

（3）超人工智能（ASI）

当前，超级智能只是一个假想的概念。人们认为将来可能会开发出这样的人工智能，但在当今世界还不存在。超级智能可以被认为处于机器超越人类能力并具有自我意识的水平。这个概念一直是多部电影和科幻小说的灵感来源，其中能够发展自己的感觉和情感的机器人可以超越人类自身。它将能够建立自己的情感，并且假设它在艺术、体育、数学、科学等方面要比人类更好。超级智能的决策能力将超过人类。超人工智能的概念对我们来说仍然是未知的，其后果也无法猜测，其影响尚无法衡量。

4. 人工智能的目标

到目前为止，你已经了解了人工智能的含义、不同级别的人工智能及其应用。但是，人工智能的目标是什么？我们旨在通过人工智能实现的结果是什么？人工智能的总体目标是使机器和计算机能够智能地学习和运行。人工智能的一些其他目标如下。

（1）解决问题

研究人员开发的算法能够模仿人类在解决难题时使用的逐步过程。在20世纪80年代后期和20世纪90年代，研究已经达到了一个新阶段，在此阶段，已经开发出处理不完整或不确定信息的方法。但是对于棘手的问题，需要巨大的计算资源和存储能力。因此，寻找能够有效解决问题的算法是人工智能的目标之一。

（2）知识表示

机器有望解决需要广泛知识的问题。因此，知识表示对于人工智能至关重要。人工智能代表对象、属性、事件、因果等。

（3）规划

人工智能的目标之一应该是设定并实现智能目标。人工智能能够就行为对变化的影响以及可用的选择做出预测。人工智能体需要评估其环境并做出预测。这就是为什么规划很重要，并且可以视为人工智能的目标。

（4）学习

人工智能的基本概念之一是机器学习，它是对计算机算法的研究，随着时间的推移，这些算法会因其经验而不断改进。机器学习有不同类型，常见的类型是无监督机器学习和监督机器学习。

（5）社会智能

情感计算本质上是对可以解释、识别和处理的人类行为的系统研究。它是计算机科学、心理学和认知科学的融合。社会智能是人工智能的另一个目标，因为在构建算法之前了解这些领域很重要。

因此，人工智能的总体目标是创建可以融合以上目标的技术，并创建可以帮助我们高效工作、更快地做出决策并提高安全性的智能机器。

5. 人工智能的未来

我们一直着迷于技术变革。当前，我们处在历史上人工智能发展最快的时期。人工智能已经在技术领域中进步最大。这不仅影响了每个行业的未来，而且还推动了大数据、机器人技术和物联网等新兴技术的发展。毫无疑问，以人工智能不断发展的速度，它将在未来继续蓬勃发展。随着人工智能及其技术的发展，在这一领域对熟练专业人员的需求将越来越大。

人工智能认证将使你从行业中的其他参与者中脱颖而出。随着面部识别、医疗保健、聊天机器人中的人工智能持续增长，现在是努力建立成功的人工智能事业的正确时机。在不知不觉间，虚拟助手已经成为我们日常生活的一部分。像特斯拉这样的科技巨头的自动驾驶汽车向我们展示了未来的前景。还有更多的进步待发现，这仅仅是开始。根据世界经济论坛的说法，到 2024 年，人工智能将创造 1.33 亿个新工作岗位。人工智能的未来绝对是光明的。

Text B

扫码听课文

Pros and Cons of Artificial Intelligence

1. Pros of artificial intelligence

In the modern days, artificial intelligence is making its way to a smarter world. AI is helping to create advanced tools for automation that can save humans' time for productive tasks. In this section , we will discuss the various pros of AI.

(1) Error-free processing

The execution of tasks by humans is more prone to make errors. We often make mistakes while doing a specific task. This might be due to the variation of the intellectual ability of an individual. But, it is not the same case with AI-based machines. We program the machines for accomplishing a specific task. Thus, the accuracy depends on how well we design and program the machines to carry out the task.

If we compare AI-based machines to humans for executing a particular task, artificial intelligence has proved itself to be more efficient than humans. The use of artificial intelligence in various fields helps reduce unnecessary errors and losses. Algorithms used for building AI-based models implement complicated mathematical constructs that help perform actions with greater efficiency and fewer errors. Thus, it helps solve complex real-world problems.

(2) Helps in repetitive jobs

Unlike humans, machines do not require breaks to recover from tiredness and boost productivity. There are many day-to-day tasks accomplished by a human, which are repetitive. The efficiency of a human reduces while continuously performing the same job. Moreover, it is a fact that a human worker can be productive only for 8–10 hours per day.

Moreover, AI-based machines can perform repetitive tasks for a long time without any slowdown. Artificial intelligence helps operate the machines for an indefinite time, without lacking productivity. This is one of the major pros of AI that have led to its acceptance in every field. Artificial intelligence is used by manufacturers to continuously produce goods to meet the market demand and earn high profits.

(3) 24×7 availability

An average worker can only invest his services for 7–8 hours per day. Humans need time to refresh themselves, and they need to maintain a work-life balance. They cannot work 24 hours a day.

Here, artificial intelligence helps provide 24×7 services to an organization. In another scenario, AI-based chatbots used by customer service applications can handle multiple queries at a time, round-the-clock. AI can provide services without any delay or lack of efficiency. Nowadays, every ecommerce application, e-learning website, healthcare sector, educational institute, etc. uses artificial intelligence for support chat. This helps in enhancing customer services.

(4) Right decision making

One of the advantages of artificial intelligence is its ability to make the right decision. There are no emotions attached to the AI-based machines, which can help prevent hampering efficiency. The machines that are built using artificial intelligence are capable of making logical decisions as well. A human would examine a situation by considering many factors. These factors may influence the decision emotionally or practically. However, the machines give accurate results as they are programmed to make logical decisions. AI-powered machines use cognitive computing that helps them make practical decisions in real time.

(5) Digital assistance

Another advantage of AI is digital assistance. AI-powered applications also provide digital assistance. Today, most of the organizations make use of digital assistants to perform automated tasks. This helps save human resources. There are digital assistants that can program to make a website for us. The use of digital assistants has revolutionized the healthcare industry as well. Now, doctors can look after their patients from remote locations with the help of digital assistants that provide real-time data of patients.

Digital assistants also help us in our day-to-day activities. There are many practical applications of AI-based digital assistants such as Google Maps, Grammarly, Alexa and many more. Google Maps helps us travel from one place to another, while Alexa executes voice searches to give us results. Another very interesting digital assistant is Grammarly that helps us correct grammar in our text, it helps auto-correct the text to improve our writing skills. These applications make artificial intelligence advantageous over other technologies.

2. Cons of AI

We have already discussed the advantages of artificial intelligence in the real world. Now let's come to its advantages.

(1) High costs of creation

The creation of machines empowered with artificial intelligence is very costly. For a large-scale project, the price might reach up to millions of dollars. Thus, for a small-scale business, it is not possible to implement AI. For companies with large revenues, the cost of the development of an AI project may be felt high due to its features, functionalities or scope with which it is designed. The cost of development also depends on the hardware and software the companies use. Moreover, to meet the demand of a highly changing world, the hardware and software should be regularly updated. AI-powered devices are built, employing complex codes, algorithms, software and hardware. The maintenance of these components requires great effort and costs very high.

(2) Increased unemployment

Artificial intelligence will create jobs as well as may leave some people unemployed. This is one of the major cons of artificial intelligence. It will create jobs for people who are skilled in technologies, but will replace low-skilled jobs. Sectors, including the manufacturing industry, have started employing AI-powered machines for manufacturing products. As AI-based machines can work 24×7, industries would prefer to invest in artificial intelligence rather than employing humans.

According to a report by Gartner, AI will create one million jobs by 2022. But, it will have a huge negative impact on low-skilled laborers as machines will replace humans there. The manufacturing industry, construction sites, the transport industry (with the advent of driverless cars), etc. will result in large-scale unemployment.

(3) Lacking creativity

Machines cannot become as creative as humans. Artificial intelligence can provide functionalities to learn from data but cannot make the machines mimic the exact human brain and skills. The accuracy of the results from an AI-powered machine depends on the level of analytics used by the creator. Artificial intelligence cannot invent anything. It can just perform the task it is programmed for and improve itself by experience.

Although AI can collaborate with other technologies such as IoT, big data, advanced sensors, etc. to give the best automation, the smartness and creativity of AI-based machines depend on how intelligent and creative the algorithms are created by humans. Therefore, AI is bound to rules and algorithms and cannot become as creative as humans.

(4) Lacking improvement

AI algorithms are designed in such a way that they allow machines to learn by themselves by exploring data. Then, the machines try to improve by learning. But, any redundancy in the data may cause failures in learning, and the machines may show unpredictable results. Then, the algorithms need to be readjusted for the new set of data or learned to adapt to exceptional conditions. Due to the lack of improvement, AI-generated results may have inaccuracy and cause great losses.

(5) No human replication

Humans have created machines to save their time and effort from doing non-essential repetitive tasks. AI-powered machines work on algorithms, mathematical computing and cognitive technologies. They can become highly advanced but cannot act or think like a human. Machines are considered intelligent, but they lack judgemental power as they are not aware of ethics, morals, the right or the wrong. The machines will break down or give unpredictable results if they find a condition for which they are not programmed. Hence, if we try to implement AI in places that require strong judgemental ability, then it would be a great failure.

There are various advantages and disadvantages of artificial intelligence. It all depends on how you want to use it.

✍ New Words

automation	[ˌɔːtəˈmeɪʃn]	n.自动化
error-free	[ˈerə friː]	adj. 无错的，无误的
mistake	[mɪˈsteɪk]	n.错误，过失 v.弄错，误解
intellectual	[ˌɪntəˈlektʃuəl]	adj.智力的；有才智的 n.知识分子；脑力劳动者
reduce	[rɪˈdjuːs]	v.减少，缩小
unnecessary	[ʌnˈnesəsəri]	adj.不需要的；没必要的
mathematical	[ˌmæθəˈmætɪkl]	adj.数学的；精确的
tiredness	[ˈtaɪədnəs]	n.疲劳，倦怠
boost	[buːst]	vt.促进，提高；增加
slowdown	[ˈsləʊdaʊn]	n.减速，减缓
operate	[ˈɒpəreɪt]	v.运转；操作
manufacturer	[ˌmænjuˈfæktʃərə]	n.制造商，制造厂
earn	[ɜːn]	v.赚得；获得
balance	[ˈbæləns]	n.均衡，平衡（能力）
Handle	[ˈhændl]	v.操作；处理 n.句柄；手柄
query	[ˈkwɪəri]	n.询问；问号 v.查询，询问
round-the-clock	[ˌraʊnd ðəˈklɒk]	adj.全天候的；不分昼夜的；连续不停的
chat	[tʃæt]	v.&n.闲谈；聊天
hamper	[ˈhæmpə]	vt.妨碍，束缚，限制
logical	[ˈlɒdʒɪkl]	adj.逻辑（上）的；符合逻辑的
examine	[ɪgˈzæmɪn]	v.检查，调查
factor	[ˈfæktə]	n.因素

influence	['ɪnfluəns]	v.影响；支配
remote	[rɪ'məʊt]	adj.远程的，遥远的
text	[tekst]	n.文本
auto-correct	['ɔːtəʊ kərekt]	v.自动校正
feature	['fiːtʃə]	n.特征，特点
		v.以……为特色
regularly	['regjələli]	adv.有规律地；经常地
unemployment	[ˌʌnɪm'plɔɪmənt]	n.失业；失业率
negative	['negətɪv]	adj.负面的，消极的
skill	[skɪl]	n.技能，技巧；本领
collaborate	[kə'læbəreɪt]	v.合作，协作
smartness	['smɑːtnəs]	n.聪明，机灵，敏捷
redundancy	[rɪ'dʌndənsi]	n.冗余，过多
unpredictable	[ˌʌnprɪ'dɪktəbl]	adj.不可预测的
readjust	[ˌriːə'dʒʌst]	v.再调整
set	[set]	n.集合；一套/副/组
exceptional	[ɪk'sepʃənl]	adj.例外的；罕见的
inaccuracy	[ɪn'ækjərəsi]	n.不准确，误差
non-essential	[ˌnɒn ɪ'senʃl]	adj.不重要的，非本质的
judgemental	[dʒʌdʒ'mentl]	adj.判断的；裁决的
moral	['mɒrəl]	n.道德
		adj.道德的

✎ Phrases

due to	由于，因为
carry out	执行；进行；完成
AI-based model	基于人工智能的模型
day-to-day task	日常任务
market demand	市场需求
customer service	客户服务
attached to	附属于
digital assistance	数字助理
human resource	人力资源
reach up to	高达
low-skilled job	低技能工作
have impact on	对……产生冲击；碰撞，影响

13

adapt to	适应
judgemental power	判断能力

Reference Translation

人工智能的利与弊

1. 人工智能的优点

在当代，人工智能正在走向一个更智能的世界。人工智能正在帮助人们创建先进的自动化工具，从而可以节省人们完成生产任务的时间。下面我们将讨论人工智能的各种优点。

（1）无错误处理

人执行任务更容易出错。我们在执行特定任务时经常会犯错误。这可能是由于个人智力的不同所致。但是，基于人工智能的机器却并非如此。我们对机器进行编程以完成特定任务。因此，完成任务的精度取决于我们对执行该任务的机器的设计和编程的程度。

就执行某一特定任务而言，如果我们将基于人工智能的机器与人类进行比较，那么人工智能已证明比人类更有效。在各个领域使用人工智能有助于减少不必要的错误和损失。用于构建基于人工智能的模型的算法可实现复杂的数学构造，从而有助于以更高的效率和更少的错误执行操作。因此，它有助于解决复杂的实际问题。

（2）帮助重复性工作

与人类不同，机器不需要休息就能从疲劳中恢复过来并提高生产率。由人类完成的许多日常任务是重复的。在连续执行相同的工作时，人的效率会降低。另一个事实是一个人每天只能有效工作 8～10 小时。

此外，基于人工智能的机器可长时间执行重复性任务，而不会降低速度。人工智能有助于无限期地操作机器，而不会缺乏生产力。这是人工智能的主要优点之一，已使其在各个领域都得到认可。制造商通过使用人工智能来连续生产商品以满足市场需求并获得高额利润。

（3）24×7 可用性

一个普通工人每天只能工作 7～8 小时。人需要时间来恢复自身的能力，他们需要保持工作与生活之间的平衡。他们无法一天工作 24 小时。

在这里，人工智能有助于为组织提供 24×7 服务。在另一种情况下，客户服务应用程序使用的基于人工智能的聊天机器人可以全天候一次处理多个查询。人工智能可以提供服务而不会造成任何延迟或效率降低。如今，每个电子商务应用程序、电子学习网站、医疗保健部门、教育机构等都使用人工智能来支持聊天。这有助于提高客户服务能力。

（4）正确的决策

人工智能的优势之一是能够做出正确的决策。基于人工智能的机器没有任何情绪，这可以帮助提高效率。使用人工智能构建的机器也能够做出合乎逻辑的决策。人在研究一种情况时会考虑许多因素。这些因素可能会在情感上或实践上影响决策。但是，因为对机器进行编程就是为了做出逻辑决策，所以它们能够给出准确的结果。人工智能驱动的机器使用认知计算来帮助它们实时做出切实可行的决策。

（5）数字助理

人工智能的另一个优势是数字助理。基于人工智能的应用程序还提供数字助理。如今，大多数组织都使用数字助理来执行自动化任务。这有助于节省人力资源。有一些数字助理可以编程为我们建立一个网站。数字助理的使用也彻底改变了医疗保健行业。现在，医生可以在为患者提供实时数据的数字助理的帮助下，实现远程诊断。

数字助理还可以帮助我们进行日常活动。基于人工智能的数字助理有许多实际应用，例如 Google Maps、Grammarly、Alexa 等。Google Maps（谷歌地图）帮助我们从一个地方旅行到另一个地方，而 Alexa 执行语音搜索为我们提供结果。Grammarly 是另一个非常有趣的数字助理，它可以帮助我们纠正文本中的语法错误，它有助于自动更正文本，以提高我们的写作技巧。这些应用使人工智能优于其他技术。

2. 人工智能的缺点

我们已经讨论了人工智能在现实世界中的优点。现在让我们来谈谈人工智能的缺点。

（1）高昂的创建成本

创建具有人工智能功能的机器的成本非常高。对于一个大型项目，价格可能高达数百万美元。因此，小型企业不可能实现人工智能。对于收入较高的公司，由于人工智能项目的特点、功能或设计范围，其开发成本可能会很高。开发成本还取决于公司使用的硬件和软件。此外，为了满足瞬息万变的世界的需求，应定期更新硬件和软件。构建人工智能驱动的设备，采用复杂的代码、算法、软件和硬件。这些组件的维护需要大量的精力，并且成本很高。

（2）增加失业

人工智能将创造就业机会，也可能使一些人失业。这是人工智能的主要弊端之一。它将为技术熟练的人们创造就业机会，但将取代低技能的工作。包括制造业在内的各个行业已开始采用人工智能驱动的机器来制造产品。由于基于人工智能的机器可以 24×7 全天候工作，因此行业更愿意在人工智能上进行投资，而不是雇用人员。

根据 Gartner 公司的报告，到 2022 年，人工智能将创造 100 万个就业机会。但是，它将对低技能劳动者产生巨大的负面影响，因为机器将替代那里的人们。导致制造业、建筑工地、运输业（随着无人驾驶汽车的问世）等行业大规模的失业。

（3）缺乏创造力

机器无法像人类一样具有创造力。人工智能具有从数据中学习的功能，但不能使机器精确地模仿人脑和技能。基于人工智能的机器的准确性取决于创建者使用的分析水平。人工智能无法创造任何东西。它只能执行为其编程的任务，并可以根据经验进行自我完善。

尽管人工智能可以与其他技术（如 IoT、大数据、高级传感器等）协作以提供最佳自动化，但是基于人工智能的机器的智能性和创造力取决于人类如何创造算法来实现智能和创造力。因此，人工智能受制于规则和算法，无法像人类一样具有创造力。

（4）缺乏改进

人工智能算法的设计方式允许机器通过探索数据自行学习。然后，机器尝试通过学习来改进。但是，数据中的任何冗余都可能导致学习失败，并且机器可能会显示不可预测的结果。然后，需要针对新的数据集重新调整算法，或者算法需要学习以适应特殊条件。由于缺乏改进，人工智能生成的结果可能不准确，并造成巨大损失。

（5）无法复制人

人类创造了机器来节省因执行不必要的重复性任务所需的时间和精力。人工智能驱动的机器可以使用算法、进行数学计算并具有认知技术。它们可以变得非常先进，但不能像人一样行动或思考。机器被认为是智能的，但它们缺乏判断力，因为它们不了解伦理、道德、正确或错误。如果发现未编程的条件，机器将崩溃或给出不可预测的结果。因此，如果我们试图在需要强大判断能力的地方实施人工智能，那将可能是一个巨大的失败。

人工智能有各种各样的优缺点，这完全取决于你要如何使用它。

Exercises

[Ex. 1] Answer the following questions according to Text A.

1. If you ask about artificial intelligence to an AI researcher, what would he say?

2. What is artificial intelligence at its core?

3. What is the goal of artificial intelligence?

4. What are the examples of weak or narrow AI?

5. What does artificial general intelligence refer to?

6. What is super-intelligence?

7. What would be the overall goal of artificial intelligence?

8. What are some of the other goals of AI?

9. What have we always been fascinated by? What has artificial intelligence emerged to be?

10. According to the World Economic Forum, how many new jobs are said to be created by artificial intelligence by the year 2024? What is the future of AI?

[Ex. 2] Fill in the following blanks according to Text B.

1. The execution of tasks by humans is more prone to _____. We often make mistakes while_____. This might be due to the variation of _____ of an individual.

2. The use of artificial intelligence in various fields helps_____ and _____. Algorithms used for building AI-based models implement complicated _____ that help perform actions _____ and _____.

3. AI-based machines can perform _____ for a long time without _____. Artificial intelligence helps operate the machines for an indefinite time, without _____.

4. An average worker can only invest his services for _____ per day while artificial intelligence helps provide _____ to an organization.

5. One of the advantages of artificial intelligence is its ability to _____. There are _____ attached to the AI-based machines, which can help prevent _____.

6.Today, most of the organizations make use of digital assistants to _____. This helps _____. Digital assistants also help us in our day-to-day activities. There are many practical applications of AI-based digital assistants such as_____, _____, _____ and many more.

7. The creation of machines empowered with _____ is _____. For a large-scale project, the price might reach up to _____. AI-powered devices are built, employing complex codes, _____, _____ and hardware. The maintenance of these components requires _____and costs very high.

8. Artificial intelligence will create jobs as well as may _____. It will create jobs for people who _____, but will replace _____.

9. Artificial intelligence can _____ to learn from data but cannot make the machines _____ and skills. The accuracy of the results from an AI-powered machine depends on _____ used by the creator.

10. AI-powered machines work on algorithms, _____ and _____. They can become highly advanced but cannot _____. Machines are considered intelligent, but they lack _____ as they are not aware of _____, _____, the right or _____.

[Ex. 3] Translate the following terms or phrases from English into Chinese.

1. ability   1. _____
2. analogy   2. _____
3. available   3. _____
4. decision   4. _____
5. entity   5. _____
6. computational resource   6. _____
7. artificial intelligence   7. _____
8. affective computing   8. _____
9. facial recognition   9. _____
10. deep learning   10. _____

[Ex. 4] Translate the following terms or phrases from Chinese into English.

1. *n.*功能；函数   1. _____
2. *n.*知识   2. _____
3. *n.*特性，属性   3. _____
4. *adj.*（计算机）虚拟的；实质上的   4. _____
5. *n.*特征，特点 *v.*以……为特色   5. _____
6. 机器智能   6. _____
7. 机器学习   7. _____
8. 语音识别   8. _____
9. 监督机器学习   9. _____
10. 虚拟助手   10. _____

[Ex. 5] Translate the following passage into Chinese.

FAQs Related to Artificial Intelligence

1. Will AI reduce jobs in future

AI is still developing. There is a huge scope for improvement and advancements in the field of AI, and although there might be some amount of upskilling required to keep up with the changing trends, AI will most likely not replace or reduce jobs in the future. In fact, a study by Gartner suggests that AI-related jobs will reach two million net-new jobs by the year 2025. The adoption of AI will help in making tasks easier for an organization.

2. How does AI work

Artificial intelligence can be built over a diverse set of components and will be merged as the following items.

- Philosophy.
- Mathematics.
- Economics.
- Neuroscience.
- Psychology.
- Computer engineering.
- Control theory and cybernetics.
- Linguistics.

3. How is artificial intelligence used in robotics

Artificial intelligence and robotics are usually seen as two different things. AI involves programming intelligence whereas robotics involves building physical robots. However, the two concepts are correlated. Robotics does use AI techniques and algorithms and AI bridges the gap between the two. Robots can be controlled by AI programs.

4. Why is artificial intelligence important

From music recommendations, map directions, mobile banking to fraud prevention, AI and other technologies have taken over. AI is important because of a number of reasons. There are several advantages to AI, such as, reduction in human error, available 24×7, helps in repetitive work, digital assistance, faster decisions, etc.

5. What are the branches of AI

Artificial Intelligence can be divided mainly into six branches. They are machine learning, neural networks, deep learning, computer vision, natural language processing and cognitive computing.

Unit 2

Text A

Knowledge Representation

Human beings are good at understanding, reasoning and interpreting knowledge. And using this knowledge, they are able to perform various actions in the real world. But how do machines perform the same? In this passage, we will learn about knowledge representation in AI and how it helps the machines perform reasoning and interpretation using artificial intelligence.

1. What is knowledge representation

Knowledge representation in AI is a study of how the beliefs, intentions and judgments of an intelligent agent can be expressed suitably for automated reasoning. One of the primary purposes of knowledge representation includes modeling intelligent behavior for an agent.

Knowledge Representation and Reasoning (KRR) represents information from the real world for a computer to understand and then utilize this knowledge to solve complex real-life problems like communicating with human beings in natural language. Knowledge representation in AI is not just about storing data in a database, it allows a machine to learn from that knowledge and behave intelligently like a human being.

The different kinds of knowledge that need to be represented in AI is as follows.

- Objects.
- Events.
- Performance.
- Facts.
- Meta-knowledge.
- Knowledge-base.

2. Different types of knowledge

There are 5 types of knowledge. They are as follows.

- Declarative knowledge: It includes concepts, facts, and objects and expressed in a declarative sentence.
- Structural knowledge: It is a basic problem-solving knowledge that describes the relationship between concepts and objects.
- Procedural knowledge: This is responsible for knowing how to do something and includes rules, strategies, procedures, etc.
- Meta-knowledge: It defines knowledge about other types of knowledge.

● Heuristic knowledge: This represents some expert knowledge in the field or subject.

3. Cycle of knowledge representation in AI

Artificial intelligent systems usually consist of various components to display their intelligent behavior. Some of these components are as follows.

- Perception.
- Learning.
- Knowledge Representation and Reasoning.
- Planning.
- Execution.

4. What is the relation between knowledge and intelligence

In the real world, knowledge plays a vital role in intelligence as well as creating artificial intelligence. It demonstrates the intelligent behavior in agents or systems. It is possible for an agent or system to act accurately on some input only when it has the knowledge or experience about the input.

5. Techniques of knowledge representation in AI

(1) Logical representation

Logical representation is a language with some definite rules which deal with propositions and has no ambiguity in representation. It represents a conclusion based on various conditions and lays down some important communication rules. Also, it consists of precisely defined syntax and semantics which supports the sound inference. Each sentence can be translated into logics using syntax and semantics, as shown in Table 2-1.

Table 2-1　Syntax and Semantics

Syntax	Semantics
● It decides how we can construct legal sentences in logic ● It determines which symbol we can use in knowledge representation ● How to write those symbols	● Semantics are the rules by which we can interpret the sentence in the logic ● It assigns a meaning to each sentence

The advantages of logical representation are as follows.

- Logical representation helps to perform logical reasoning.
- This representation is the basis for the programming languages.

The disadvantages of logical representation are as follows.

- Logical representation has some restrictions and is challenging to work with.
- This technique may not be very natural, and inference may not be very efficient.

(2) Semantic network representation

Semantic networks work as an alternative of predicate logic for knowledge representation. In semantic networks, you can represent your knowledge in the form of graphical networks. This network consists of nodes representing objects and arcs which describe the relationship between those objects. Also, it categorizes the object in different forms and links those objects.

This representation consist of two types of relations.

- IS-A relation (Inheritance).
- Kind-of-relation.

The advantages of semantic network representation are as follows.

- Semantic networks are a natural representation of knowledge.
- Also, it conveys meaning in a transparent manner.
- Semantic networks are simple and easy to understand.

The disadvantages of semantic network representation are as follows.

- Semantic networks take more computational time at runtime.
- Also, these are inadequate as they do not have any equivalent quantifiers.
- Semantic networks are not intelligent and depend on the creator of the system.

(3) Frame representation

A frame is a record like structure that consists of a collection of attributes and values to describe an entity in the world. The top layer represents a fixed concept, object or event. The other layers are composed of some structures called slots. Slots have slot names and values, or have value constraints.

The advantages of frame representation are as follows.

- It makes the programming easier by grouping the related data.
- Frame representation is easy to understand and visualize.
- It is very easy to add slots for new attributes and relations.
- Also, it is easy to include default data and search for missing values.
- It has a very generalized approach.

The disadvantages of frame representation are as follows.

- In frame system inference, the mechanism cannot be easily processed.
- The inference mechanism cannot be smoothly proceeded by frame representation.

(4) Production rules

In production rules, the agent checks for the condition and if the condition exists then production rule fires and corresponding action is carried out. The condition part of the rule determines which rule may be applied to a problem. The action part carries out the associated problem-solving steps. This complete process is called a recognize - act cycle.

The production rules system consists of three main parts.

- The set of production rules.
- Working memory.
- The recognize-act cycle.

The advantages of production rules are as follows.

- The production rules are expressed in natural language.
- The production rules are highly modular and can be easily removed or modified.

The disadvantages of production rules are as follows.

- It does not exhibit any learning capabilities or store the result of the problem for future uses.

- During the execution of the program, many rules may be active. Thus, rule-based production systems are inefficient.

6. Representation requirements

A good knowledge representation system must have the following properties.

- Representational accuracy: It should represent all kinds of required knowledge.
- Inferential adequacy: It should be able to manipulate the representational structures to produce new knowledge corresponding to the existing structure.
- Inferential efficiency: The ability to direct the inferential knowledge mechanism into the most productive directions by storing appropriate guides.
- Acquisitional efficiency: The ability to acquire new knowledge easily using automatic methods.

7. Approaches to knowledge representation in AI

There are different approaches to knowledge representation.

(1) Simple relational knowledge

It is the simplest way of storing facts which uses the relational method. Here, all the facts about a set of the object are set out systematically in columns. Also, this approach of knowledge representation is famous in database systems where the relationship between different entities is represented. Thus, there is little opportunity for inference.

Example is shown in Table 2-2.

Table 2-2　An example of representing simple relational knowledge

Name	Age	Emp ID
John	25	100071
Amanda	23	100056
Sam	27	100042

This is an example of representing simple relational knowledge.

(2) Inheritable knowledge

In the inheritable knowledge approach, all data must be stored into a hierarchy of classes and should be arranged in a generalized form or a hierarchal manner. Also, this approach contains inheritable knowledge which shows a relation between instance and class, and it is called instance relation. In this approach, objects and values are represented in Boxed nodes.

(3) Inferential knowledge

The inferential knowledge approach represents knowledge in the form of formal logic. Thus, it can be used to derive more facts. Also, it guarantees correctness.

✎ New Words

representation	[ˌreprɪzen'teɪʃn]	n.表示，表现；陈述
interpret	[ɪn'tɜ:prət]	v.解释，诠释；领会

interpretation	[ɪnˌtɜ:prəˈteɪʃn]	n.解释，说明
judgment	[ˈdʒʌdʒmənt]	n.判断，鉴定；辨别力，判断力
agent	[ˈeɪdʒənt]	n.实体；代理
suitably	[ˈsu:təbli]	adv.适当地，适宜地；相配地；合适地
database	[ˈdeɪtəbeɪs]	n.数据库
object	[ˈɒbdʒɪkt]	n.对象；物体；目标
event	[ɪˈvent]	n.事件
performance	[pəˈfɔ:məns]	n.性能；执行
meta-knowledge	[ˌmetə ˈnɒlɪdʒ]	n.元知识
knowledge-base	[ˈnɒlɪdʒ beɪs]	n.知识库
declarative	[dɪˈklærətɪv]	adj.陈述的
structural	[ˈstrʌktʃərəl]	adj.结构（上）的
procedural	[prəˈsi:dʒərəl]	adj.程序的；过程的
rule	[ru:l]	n.规则 v.控制，支配
heuristic	[hjuˈrɪstɪk]	adj.启发式的
perception	[pəˈsepʃn]	n.洞察力；知觉
execution	[ˌeksɪˈkju:ʃn]	n.实施，执行
demonstrate	[ˈdemənstreɪt]	v.证明；说明；演示
definite	[ˈdefɪnət]	adj.不会改变的；明确的
semantic	[sɪˈmæntɪk]	adj.语义的，语义学的
construct	[kənˈstrʌkt]	v.构造；组成
symbol	[ˈsɪmbl]	n.符号，记号
restriction	[rɪˈstrɪkʃn]	n.限制，限定
categorize	[ˈkætəgəraɪz]	vt.把……归类，把……分门别类
inheritance	[ɪnˈherɪtəns]	n.继承
convey	[kənˈveɪ]	v.表达；传送
runtime	[ˈrʌntaɪm]	n.运行期，运行时间
inadequate	[ɪnˈædɪkwət]	adj.不充足的；不适当的
equivalent	[ɪˈkwɪvələnt]	adj.相等的，相同的 n.等同物；对应物
quantifier	[ˈkwɒntɪfaɪə]	n.量词
attribute	[əˈtrɪbju:t]	n.属性，性质，特征
slot	[slɒt]	n.槽
mechanism	[ˈmekənɪzəm]	n.机制
smoothly	[ˈsmu:ðli]	adv.平滑地；流畅地；平稳地

modify	['mɒdɪfaɪ]	v.修改，改变
exhibit	[ɪg'zɪbɪt]	v.表现；展览
inefficient	[ˌɪnɪ'fɪʃnt]	adj.无效率的，无能的
systematically	[ˌsɪstə'mætɪkli]	adv.有系统地；有组织地
inheritable	[ɪn'herɪtəbl]	adj.可继承的，会遗传的
instance	['ɪnstəns]	n.实例
guarantee	[ˌgærən'tiː]	v.确保
		n.保证；保修单
correctness	[kə'rektnəs]	n.正确性

✍ Phrases

intelligent agent	智能体；智能代理
automated reasoning	自动推理
natural language	自然语言
declarative knowledge	陈述性知识
structural knowledge	结构化知识
procedural knowledge	程序性知识
heuristic knowledge	启发式知识
logical reasoning	逻辑推理
programming language	程序设计语言，编程语言
semantic network	语义网
predicate logic	谓词逻辑
graphical network	图形化的网络
frame representation	框架表示
production rule	产生式规则
recognize-act cycle	识别-行动循环
relational method	关系法
set out	列出；安排；摆放；陈列
inheritable knowledge	可继承的知识
instance relation	实例关系
inferential knowledge	推理知识
formal logic	形式逻辑

✍ Abbreviations

| KRR (Knowledge Representation and Reasoning) | 知识表示和推理 |

Reference Translation

知识表示

人类善于理解、推理和解释知识。利用这些知识，他们能够在现实世界中执行各种动作。但是机器如何执行相同的操作？在本文中，我们将学习 AI 中的知识表示以及它如何帮助机器使用人工智能来进行推理和解释。

1. 什么是知识表示

AI 中的知识表示是关于如何适当地表达智能体的信念、意图和判断，以进行自动推理的研究。知识表示的主要目的之一是为智能体建立智能行为模型。

知识表示和推理（KRR）将来自现实世界的信息表示出来供计算机理解，然后利用这些知识来解决复杂的现实生活问题，如用自然语言与人交流。AI 中的知识表示不仅是将数据存储在数据库中，还使机器可以从该知识中学习，并像人类一样智能地运行。

AI 中需要表达的各种知识如下。

- 对象。
- 事件。
- 表现。
- 事实。
- 元知识。
- 知识库。

2. 不同类型的知识

有 5 种类型的知识，它们分别如下。

- 陈述性知识：它包括概念、事实和对象，并用陈述句表达。
- 结构化知识：它是解决问题的基本知识，描述了概念和对象之间的关系。
- 程序性知识：负责了解如何做某事，包括规则、策略及过程等。
- 元知识：它定义有关其他类型知识的知识。
- 启发式知识：表示该领域或主题中的一些专家知识。

3. 人工智能中的知识表示周期

人工智能系统通常由各个部分组成，以显示其智能行为。其中一些组件如下。

- 洞察力。
- 学习。
- 知识表示与推理。
- 规划。
- 执行。

4. 知识与智力之间有什么关系

在现实世界中，知识在智能以及创建人工智能中起着至关重要的作用。它展示了智能体或系统中的智能行为。仅当智能体或系统具有相关输入的知识或经验时，才有可能对某些输入进行准确的操作。

5. 人工智能中的知识表示技术

（1）逻辑表示

逻辑表示是一种具有确定规则的语言，用于处理命题，并且在表示上没有歧义。它代表了基于各种条件的结论，并规定了一些重要的通信规则。而且，它由精确定义的语法和语义组成，这些语法和语义支持合理推断。可以使用语法和语义将每个句子变成逻辑表达，见表2-1。

表2-1 语法和语义

语　　法	语　　义
● 它决定了我们如何构建逻辑上的合法语句 ● 它确定我们可以在知识表示中使用哪些符号 ● 如何写这些符号	● 语义是我们按逻辑来解释句子的规则 ● 它为每个句子分配了含义

逻辑表示的优点如下。

● 逻辑表示有助于执行逻辑推理。

● 这种表示形式是编程语言的基础。

逻辑表示的缺点如下。

● 逻辑表示形式有一些限制，使用起来有挑战性。

● 这种技术可能不是很自然，并且推理可能不是很有效。

（2）语义网络表示

语义网络是知识表示的谓词逻辑的替代方法。在语义网络中，你可以用图形网络的形式表示你的知识。该网络由代表对象的节点和描述这些对象之间关系的弧组成。而且，它以不同的形式对对象进行分类并链接这些对象。

此表示形式包含两种类型的关系。

● IS-A 关系（继承）。

● 关系类型。

语义网络表示的优点如下。

● 语义网络是知识的自然表示。

● 此外，它以透明的方式传达含义。

● 语义网络简单易懂。

语义网络表示的缺点如下。

● 语义网络在运行时需要花费更多的计算时间。

● 另外，只有语义网络还不够，因为它们没有任何等效的量词。

● 语义网络不是智能的，并且取决于系统的创建者。

（3）框架表示

框架是一种类似于记录的结构，由一组属性和值组成，用以描述世界上的实体。顶层代表一个固定的概念、对象或事件。其他层由一些称为槽的结构组成。槽具有名称和值，或者具有值约束。

框架表示的优点如下。

● 通过对相关数据进行分组，使编程变得更加容易。

● 框架表示易于理解和可视化。

- 为新属性和关系添加槽非常容易。
- 此外，很容易包含默认数据并搜索缺失值。
- 它具有非常通用的方法。

框架表示的缺点如下。

- 在框架系统推理中，无法轻松处理该机制。
- 不能通过框架表示来顺利地进行推理机制。

（4）产生式规则

在产生式规则中，智能体检查条件，如果条件存在，则触发产生式规则并执行相应的操作。规则的条件部分确定可以应用于问题的规则。行动部分执行相关问题的解决步骤。这个完整的过程称为识别-行动循环。

产生式规则系统包括三个主要部分。

- 产生式规则集。
- 工作内存。
- 识别-行动循环。

产生式规则的优点如下。

- 产生式规则用自然语言表达。
- 产生式规则高度模块化，可以轻松删除或修改。

产生式规则的缺点如下。

- 它不具有任何学习功能，也不存储问题的结果以备将来使用。
- 在程序执行期间，许多规则可能处于活动状态。因此，基于规则的生产系统效率低下。

6. 表示要求

良好的知识表示系统必须具有以下特性。

- 表示精确：它应该表示所有必需的知识。
- 推论充足：它应该能够操控表示结构，以产生与现有结构相对应的新知识。
- 推理有效：通过存储适当的指导方法，能将推理知识机制引导到最有效的方向。
- 习得有效：具有使用自动方法轻松掌握新知识的能力。

7. 人工智能中的知识表示方法

知识表示有不同的方法。

（1）简单的关系知识

这是使用关系方法存储事实的最简单方法。在这里，有关一组对象的所有事实都在各栏中系统地列出。同样，这种知识表示方法在表示不同实体之间关系的数据库系统中也很有用。因此，几乎没有推理的机会。

示例见表2-2。

表2-2　一个简单关系知识的示例

姓名	年龄	Emp ID
John	25	100071
Amanda	23	100056
Sam	27	100042

这是表示简单关系知识的示例。

（2）可继承的知识

在可继承知识方法中，所有数据必须存储在类的层次结构中，并应以广义形式或层次结构方式进行排列。另外，这种方法包含可继承的知识，该知识显示了实例与类之间的关系，称为实例关系。在这种方法中，对象和值在装箱节点中表示。

（3）推理知识

推理知识方法以形式逻辑的形式表示知识。因此，它可以用来推导出更多的事实。而且，它还保证了正确性。

Text B

What Is a Knowledge Graph

扫码听课文

Knowledge graph is a model of a knowledge domain created by experts with the help of intelligent machine learning algorithms. It provides a structure and common interface for all the data and enables the creation of smart multilateral relations throughout the databases. Structured as an additional virtual data layer, the knowledge graph lies on top of the existing databases or data sets to link all the data together at scale—be it structured or unstructured.

1. Key characteristics

Knowledge graph combine characteristics of several data management paradigms.

- Database, because the data can be explored via structured queries.
- Graph, because they can be analyzed as any other network data structure.
- Knowledge base, because they bear formal semantics, which can be used to interpret the data and infer new facts.

Knowledge graph represented in RDF provide the best framework for data integration, unification, linking and reuse, because they combine the following features.

- Expressivity: The standards in the semantic web stack (RDF and OWL) allow for a fluent representation of various types of data and content, such as data schema, taxonomies and vocabularies, all sorts of metadata, reference and master data.
- Performance: All the specifications have been thought out and proven in practice, to allow for efficient management of graphs of billions of facts and properties.
- Interoperability: There is a range of specifications for data serialization, access, management and federation. The use of globally unique identifiers facilitates data integration and publishing.
- Standardization: All the above is standardized through the W3C community process, to make sure that the requirements of different actors are satisfied.

2. Ontologies and formal semantics

Ontologies represent the backbone of the formal semantics of a knowledge graph. They can be seen as the data schema of the graph. They serve as a formal contract between the developers of the knowledge graph and its users regarding the meaning of the data in it. A user could be another human being or a software application that wants to interpret the data in a reliable and precise way. Ontologies ensure a shared understanding of the data and its meanings.

When formal semantics are used to express and interpret the data of a knowledge graph, there are a number of representation and modeling instruments.

(1) Classes

Most often an entity description contains a classification of the entity with respect to a class hierarchy. For instance, when dealing with business information there could be classes Person, Organization and Location. Persons and Organization can have a common superclass Agent. Location usually has numerous subclasses, e.g., Country, Populated place, City, etc. The notion of class is borrowed by the object-oriented design, where each entity usually belongs to exactly one class.

(2) Relationship types

The relationships between entities are usually tagged with types, which provide information about the nature of the relationship, e.g., friend, relative, competitor, etc. Relationship types can also have formal definitions, e.g., that parent-of is inverse relation of child-of, they both are special cases of relative-of, which is a symmetric relationship. Or defining that sub-region and subsidiary are transitive relationships.

(3) Categories

An entity can be associated with categories, which describe some aspect of its semantics, e.g., "Big four consultants" or "XIX century composers". A book can belong simultaneously to all these categories: "Books about Africa" "Bestseller" "Books by Italian authors" "Books for kids", etc. The categories are described and ordered into taxonomy.

(4) Free text descriptions

Often a "human-friendly text" description is provided to further clarify design intentions for the entity and improve search.

3. What is not a knowledge graph

Not every RDF graph is a knowledge graph. For instance, a set of statistical data, e.g. the GDP data for countries, represented in RDF is not a KG (Knowledge Graph). A graph representation of data is often useful, but it might be unnecessary to capture the semantic knowledge of the data. It's the connections and the graph that make the KG , not the language used to represent the data.

Not every knowledge base is a knowledge graph. A key feature of a KG is that entity descriptions should be interlinked to one another. The definition of one entity includes another entity. This linking is how the graph forms (e.g. A is B, B is C, C has D, A has D). Knowledge bases without formal structure and semantics, e.g. Q&A "knowledge base" about a software product, also do not represent a KG. It is possible to have an expert system that has a collection of data organized in a format that is not a graph but uses automated deductive processes such as a set of "if-then" rules to facilitate analysis.

4. Examples of big knowledge graph

(1) Google knowledge graph

Google made this term popular with the announcement of its knowledge graph in 2012.

However, there are very few technical details about its organization, coverage and size. There are also very limited means for using this knowledge graph outside Google's own projects.

(2) DBPedia

This project leverages the structure inherent in the infoboxes of Wikipedia to create an enormous dataset (link is https://wiki.dbpedia.org/about) and an ontology that has encyclopedic coverage of entities such as people, places, films, books, organizations, species, diseases, etc. This dataset is at the heart of the Open Linked Data movement. It has been invaluable for organizations to bootstrap their internal knowledge graphs with millions of crowdsourced entities.

(3) GeoNames

Under a creative commons, users of GeoNames dataset have access to 25 million geographical entities and features.

(4) WordNet

One of the most well-known lexical databases for the English language, providing definitions and synonyms. Often used to enhance the performance of NLP and search applications.

(5) FactForge

After years of developing expertise in the news publishing industry, Ontotext produced their knowledge graph of Linked Open Data and news articles about people, organizations and locations. It incorporates the data from the KG described above as well as specialized ontologies such as the Financial Industry Business Ontology.

5. Knowledge graph and RDF databases

Years ago, we moved away from the buzzword of Big Data to Smart Data. Having unprecedented amounts of data pushed the need to have a data model that mirrored our own understanding of complex information.

To make data smart, the machines needed to be no longer bound by inflexible data schemas. We needed data repositories that could represent the "real world" and the tangled relationships that are entailed. All this needed to be done in a machine-readable way and have a formal semantics to enable automated reasoning that complemented and facilitated our own.

RDF databases (also called RDF triplestores), such as Ontotext's GraphDB, can smoothly integrate heterogeneous data from multiple sources and store hundreds of billions of facts. The RDF graph structure is very robust (it can handle massive amounts of data of all kinds and from various sources) and flexible (it does not need its schema re-defined every time we add new data).

As we have already seen, there are many freely available interlinked facts from sources such as DBpedia, GeoNames, Wikidata and so on, and their number continues to grow every day. However, the real power of knowledge graph comes when we transform our own data into RDF triples and then connect our proprietary knowledge to open global knowledge.

Another important feature of RDF databases is their inference capability where new knowledge can be created from already existing facts. When such new facts are materialized and stored in an RDF database, our search results become much more relevant, opening new avenues for actionable insights.

But if we want to add even more power to our data, we can use text mining techniques to extract the important facts from free-flowing texts and then add them to the facts in our database.

6. How can knowledge graph help text analysis

It is no surprise that modern text analysis technology makes considerable use of knowledge graphs.

- Big graphs provide background knowledge, human-like concept and entity awareness, to enable a more accurate interpretation of the text.
- The results of the analysis are semantic tags (annotations) that link references in the text to specific concepts in the graph. These tags represent structured metadata that enables better search and further analytics.
- Facts extracted from the text can be added to enrich the knowledge graph, which makes it much more valuable for analysis, visualization and reporting.

Ontotext platform implements all flavors of this interplay linking text and big knowledge graphs to enable solutions for content tagging, classification and recommendation. It is a platform for organizing enterprise knowledge into knowledge graphs, which consists of a set of databases, machine learning algorithms, APIs and tools for building various solutions for specific enterprise needs.

7. What are knowledge graph used for

A number of specific uses and applications rely on knowledge graph. Examples include data and information-heavy services such as intelligent content and package reuse, responsive and contextually aware content recommendation, knowledge graph powered drug discovery, semantic search, investment market intelligence, information discovery in regulatory documents, advanced drug safety analytics, etc.

🖎 New Words

multilateral	[ˌmʌltiˈlætərəl]	*adj.*多方面的，多边的
link	[lɪŋk]	*v.*连接 *n.*超文本链接
bear	[beə]	*v.*带有，拥有
unification	[ˌjuːnɪfɪˈkeɪʃn]	*n.*统一，联合；一致
reuse	[ˌriːˈjuːz]	*vt.*复用，重用
expressivity	[ˌekspreˈsɪvəti]	*n.*表现力，表达性
fluent	[ˈfluːənt]	*adj.*流畅的；流利的
metadata	[ˈmetədeɪtə]	*n.*元数据
prove	[pruːv]	*v.*验证，证实
interoperability	[ˌɪntərˌɒpərəˈbɪləti]	*n.*互用性，协同工作的能力
federation	[ˌfedəˈreɪʃn]	*n.*联邦；同盟，联合会
standardization	[ˌstændədaɪˈzeɪʃn]	*n.*标准化；规格化；规范化

satisfy	['sætɪsfaɪ]	v.（使）满意，满足
ontology	[ɒn'tɒlədʒi]	n.本体论，实体论
class	[klɑːs]	n.类
superclass	['sjuːpəklɑːs]	n.超类
subclass	['sʌbklɑːs]	n.子类
borrow	['bɒrəʊ]	v.借用，引用
sub-region	[sʌb 'riːdʒən]	n.子域
transitive	['trænzətɪv]	adj.传递的
taxonomy	[tæk'sɒnəmi]	n.分类法，分类学，分类系统
capture	['kæptʃə]	v.捕捉，捕获
coverage	['kʌvərɪdʒ]	n.覆盖范围
infobox	['ɪnfəʊbɒks]	n.信息框
encyclopedic	[ɪnˌsaɪklə'piːdɪk]	adj.百科全书的
bootstrap	['buːtstræp]	n.引导程序
crowdsource	[kraʊdsɔːs]	vt.众包
geographical	[ˌdʒiːə'ɡræfɪkl]	adj.地理的
mirror	['mɪrə]	vt.反映；反射
tangle	['tæŋɡl]	n.纠缠；混乱 v.（使）缠结，（使）乱作一团
machine-readable	[məˈʃiːn 'riːdəbl]	adj.机器可读的，可用计算机处理的
triplestore	['trɪplstɔː]	n.三元存储
heterogeneous	[ˌhetərə'dʒiːniəs]	adj.异构的，各种各样的
triple	['trɪpl]	n.三元组
materialize	[mə'tɪəriəlaɪz]	vi.具体化；实质化
awareness	[ə'weənəs]	n.意识；了解；觉察
annotation	[ˌænə'teɪʃn]	n.注释
enrich	[ɪn'rɪtʃ]	v.使充实；使丰富
regulatory	['reɡjələtəri]	adj.监管的

✍ Phrases

knowledge domain	知识领域
common interface	通用接口
virtual data layer	虚拟数据层
data management paradigm	数据管理范式
structured query	结构化查询
network data structure	网络数据结构

formal semantic	形式语义
data schema	数据模式
master data	主数据
data serialization	数据序列化
globally unique identifier	全局唯一标识符
serve as	充当，担任
shared understanding	共识
modeling instrument	建模工具
object-oriented design	面向对象设计
special case	特例
symmetric relationship	对称关系
graph representation	图形表示
news publishing industry	新闻出版行业
heterogeneous data	异构数据
text mining technique	文本挖掘技术
text analysis technology	文本分析技术
semantic tag	语义标记
information-heavy service	信息密集型服务

✍ Abbreviations

RDF (Resource Description Framework)	资源描述框架
OWL (Web Ontology Language)	网络本体语言
W3C(World Wide Web Consortium)	万维网联盟，又称 W3C 理事会
GDP (Gross Domestic Product)	国内生产总值
KG (Knowledge Graph)	知识图谱

Reference Translation

什么是知识图谱

知识图谱是专家在智能机器学习算法的帮助下创建的知识领域的模型。它为所有数据提供结构和通用接口，并允许在整个数据库中创建智能多边关系。知识图谱被结构化为一个附加的虚拟数据层，位于现有数据库或数据集的顶部，可以按比例将所有数据链接在一起——无论是结构化的还是非结构化的。

1. 主要特点

知识图谱结合了几种数据管理范例的特征。

● 数据库，因为可以通过结构化查询来仔细查看数据。

- 图谱，因为它们可以像其他任何网络数据结构一样进行分析。
- 知识库，因为它们具有形式语义，可用于解释数据和推断新事实。

用 RDF 表示的知识图谱为数据集成、统一、链接和重用提供了最佳框架，因为它们结合了以下特性。

- 表现力：语义网堆栈中的标准（RDF 和 OWL）可以流畅地表示各种类型的数据和内容，如数据模式、分类法和词汇表、各种元数据、参考数据和主数据。
- 性能：所有规范都经过深思熟虑并在实践中得到证明，可以高效管理数十亿个事实和属性的图谱。
- 互操作性：具有数据序列化、访问、管理和联合的一系列规范。全局唯一标识符的使用有助于数据集成和发布。
- 标准化：以上所有内容都是通过 W3C 社区过程进行标准化的，以确保满足不同参与者的要求。

2. 本体与形式语义学

本体代表知识图谱形式语义的骨干。它们可以看作是图的数据模式。它们充当知识图谱的开发者与用户之间关于其中数据含义的正式契约。用户可以是另一个人，也可以是想要以可靠且精确的方式解释数据的软件应用程序。本体确保对数据及其含义的共识。

当使用形式语义来表达和解释知识图谱的数据时，有许多表示和建模工具。

（1）类

大多数情况下，实体描述包含关于类层次结构的实体分类。例如，在处理业务信息时，可能会有人员、组织和位置类别。人员和组织可以具有共同的超类代理。位置通常有许多子类，如国家、人口稠密的地方、城市等。面向对象设计借用了类的概念，其中每个实体通常完全属于一个类。

（2）关系类型

实体之间的关系通常用类型标记，这些类型提供有关关系性质的信息，如朋友、亲戚、竞争者等。关系类型也可以具有形式定义，例如，"父代"是"子代"的逆向关系，它们都是相对关系的特例，这是一种对称关系。或将子区域和子公司定义为传递关系。

（3）类别

实体可以与描述其语义某些方面的类别相关联，例如"四大顾问"或"十九世纪作曲家"。一本书可以同时属于所有这些类别："关于非洲的书籍""畅销书""意大利作家的书籍""儿童书籍"等。这些类别按分类法进行描述和排序。

（4）自由文本描述

通常会提供"人性化的文字"描述，以进一步阐明实体的设计意图并改善搜索范围。

3. 什么不是知识图谱

并非每个 RDF 图都是知识图谱。例如，一组统计数据，以 RDF 表示的国家/地区的 GDP 数据不是 KG（知识图谱）。数据的图形表示通常很有用，但可能不必捕获数据的语义知识。构成 KG 的是连接和图形，而不是用来表示数据的语言。

并非每个知识库都是知识图谱。KG 的一项关键特性是实体描述应相互链接。一个实体的定义包括另一个实体。这种链接是图形的形成方式（例如，A 为 B，B 为 C，C 具有 D，A 具有 D）。没有形式结构和语义的知识库，如关于软件产品的问答"知识库"也并不是

KG。可能会有一个专家系统，该系统有一个以非图形格式组织的数据集合，但是使用自动演绎过程（如"如果-则"规则集）来促进分析。

4. 大知识图谱的例子

（1）Google 知识图谱

Google 于 2012 年发布了知识图谱，从而使该术语大受欢迎。但是，有关其组织、覆盖范围和规模的技术细节却很少。在 Google 自己的项目之外使用这种知识图谱的方法也非常有限。

（2）DBPedia

该项目利用了 Wikipedia 信息框中固有的结构来创建一个巨大的数据集（链接为 https://wiki.dbpedia.org/about）和一个本体，该本体像百科全书一样，涵盖了许多实体，诸如人、地方、电影、书籍、组织、物种、疾病等。此数据集是"开放链接数据"运动的核心。对于组织而言，利用数百万众包实体来引导其内部知识图谱价值极高。

（3）GeoNames

在知识共享下，GeoNames 数据集的用户可以访问 2500 万个地理实体和特征。

（4）WordNet

英语最著名的词汇数据库之一，提供定义和同义词。通常用于增强 NLP 和搜索应用程序的性能。

（5）FactForge

经过多年在新闻出版行业的专业知识开发，Ontotext 制作了它们的知识图谱，其中包括有关人员、组织和地点的链接开放数据与新闻文章。它结合了来自上述 KG 的数据以及专门的本体，如金融行业业务本体。

5. 知识图谱和 RDF 数据库

几年前，大数据这个流行词转移到了智能数据。由于拥有了前所未有的数据量，就迫切需要建立一个能够反映我们对复杂信息的理解的数据模型。

为了使数据变得智能，机器不再需要被不灵活的数据模式所约束。我们需要可以代表"现实世界"以及由此产生的复杂关系的数据存储库。所有这些都需要以机器可读的方式完成，并具有形式语义以实现自动推理，从而补充和促进我们自己的推理。

RDF 数据库（也称为 RDF 三元存储），如 Ontotext 的 GraphDB，可以平稳地集成来自多个来源的异构数据，并存储数千亿的事实。RDF 图结构非常健壮（它可以处理各种来源的大量数据）和灵活（不需要在每次添加新数据时都重新定义其架构）。

正如我们已经看到的，从 DBpedia、GeoNames、Wikidata 等来源可以免费获得相互关联的事实，并且它们的数量每天都在增长。但是，当我们将自己的数据转换为 RDF 三元组，然后将我们的专有知识连接到开放的全球知识时，知识图谱的真正力量就会显现。

RDF 数据库的另一个重要特性是其推理能力，即可以根据现有事实创建新知识。当这些新事实被具体化并存储在 RDF 数据库中时，我们的搜索结果将变得更加相关，从而为可操作的见解开辟了新途径。

但是，如果我们想为数据增加更多的功能，则可以使用文本挖掘技术从自由流动的文本中提取重要事实，然后将其添加到数据库的事实中。

6. 知识图谱如何帮助文本分析

现代文本分析技术大量使用知识图谱不足为奇。

- 大的图谱提供背景知识、类人概念和实体意识，从而可以更准确地解释文本。
- 分析的结果是语义标记（注释），这些标记将文本中的引用链接到图形中的特定概念。这些标记表示结构化的元数据，可以更好地进行搜索和进一步分析。
- 可以添加从文本中提取的事实以丰富知识图谱，这对于分析、可视化和报告来说具有更大的价值。

Ontotext 平台实现了这种相互作用的链接文本和大型知识图谱的所有形式，从而为内容标记、分类和推荐提供了解决方案。这是一个将企业知识组织成知识图谱的平台，该平台由一组数据库、机器学习算法、API 和工具组成，用于为特定企业需求构建各种解决方案。

7. 知识图谱有什么用

许多特定的用途和应用都依赖于知识图谱。示例包括数据和信息密集型服务，如智能内容和包装重用、响应和上下文感知的内容推荐、知识图谱支持的药物发现、语义搜索、投资市场情报、监管文件中的信息发现、先进的药物安全分析等。

Exercises

[Ex. 1] Answer the following questions according to Text A.

1.What is knowledge representation basically?

2. What do the different kinds of knowledge that need to be represented in AI include?

3. How many types of knowledge are there? What are they?

4. Artificial intelligent systems usually consist of various components to display their intelligent behavior. What do some of these components include?

5. What is logical representation?

6. What does semantic network consist of?

7. What is a frame?

8. What does the production rules system consist of?

9. What properties must a good knowledge representation system have?

10. What are the different approaches to knowledge representation?

[Ex. 2] Answer the following questions according to Text B.

1. What is knowledge graph?

2. What does knowledge graph represented in RDF provide? Why?

3. What do ontologies serve as?

4. What are the representation and modeling instruments mentioned in the passage?

5. What are the relationships between entities usually tagged with? What do they provide?

6. What is a key feature of a KG?

7. What does DBPedia do?

8. What is WordNet?

9. What is another important feature of RDF databases?

10. What is Ontotext platform?

[Ex. 3] **Translate the following terms or phrases from English into Chinese.**

1. agent
2. attribute
3. database
4. event
5. heuristic
6. automated reasoning
7. formal logic
8. inferential knowledge
9. intelligent agent
10. logical reasoning

1. _____
2. _____
3. _____
4. _____
5. _____
6. _____
7. _____
8. _____
9. _____
10. _____

[Ex. 4] **Translate the following terms or phrases from Chinese into English.**

1. *n.*元知识
2. *n.*对象；物体；目标的
3. *adj.*程序的；过程的
4. *n.*运行期，运行时间
5. *n.*意识；了解；觉察
6. 知识表示
7. 产生式规则
8. 通用接口
9. 数据模式
10. 面向对象设计

1. _____
2. _____
3. _____
4. _____
5. _____
6. _____
7. _____
8. _____
9. _____
10. _____

[Ex. 5] Translate the following passage into Chinese.

Knowledge Base

A knowledge base is a database used for knowledge sharing and management. It promotes the collection, organization and retrieval of knowledge. Many knowledge bases are structured around artificial intelligence and not only store data but find solutions for further problems using data from previous experience stored as part of the knowledge base.

Knowledge management systems depend on data management technologies ranging from relational databases to data warehouses. Some knowledge bases are little more than indexed encyclopedic information; others are interactive and behave/respond according to the input prompted from the user.

A knowledge base is not merely a space for data storage, but can be an artificial intelligence tool for delivering intelligent decisions. Various knowledge representation techniques including frames and scripts, represent knowledge. The services offered are explanation, reasoning and intelligent decision support.

Knowledge-Based Computer-Aided Systems Engineering (KB-CASE) tools assist designers by providing suggestions and solutions, thereby helping to investigate the results of design decisions. The knowledge base analysis and design allows users to frame knowledge bases, from which informative decisions are made.

The two major types of knowledge bases are human readable and machine readable.

- Human readable knowledge bases enable people to access and use the knowledge. They store help documents, manuals, troubleshooting information and frequently answered questions. They can be interactive and lead users to solutions to problems they have, but rely on the user providing information to guide the process.

- Machine readable knowledge bases store knowledge, but only in system readable forms. Solutions are offered based upon automated deductive reasoning and are not so interactive as this relies on query systems that have software that can respond to the knowledge base to narrow down a solution. This means that machine readable knowledge base information shared with other machines is usually linear and is limited in interactivity, unlike the human interaction which is query based.

Knowledge Management (KM) contains a range of strategies used in an organization to create, represent, analyze, distribute and enable the adoption of experiences. It focuses on competitive advantages and the improved performance of organizations.

Unit 3

Text A

Reasoning in Artificial Intelligence

We have learned various ways of knowledge representation in artificial intelligence. Now let's come to the various ways to reason on this knowledge using different logical schemes.

1. Reasoning

Reasoning is the mental process of deriving logical conclusion and making predictions from available knowledge, facts and beliefs. Or we can say, "Reasoning is a way to infer facts from existing data." It is a general process of thinking rationally to find valid conclusions.

In artificial intelligence, reasoning is essential so that the machine can also think rationally as a human brain and can perform like a human.

2. Types of reasoning

In artificial intelligence, reasoning can be divided into the following categories: deductive reasoning, inductive reasoning, abductive reasoning, common sense reasoning, monotonic reasoning, non-monotonic reasoning.

Note: Inductive and deductive reasoning are the forms of propositional logic.

(1) Deductive reasoning

Deductive reasoning is deducing new information from logically related known information. It is the form of valid reasoning, which means the argument's conclusion must be true when the premises are true.

Deductive reasoning is a type of propositional logic in AI and it requires various rules and facts. It is sometimes referred to as top-down reasoning and it is contradictory to inductive reasoning.

In deductive reasoning, the truth of the premises guarantees the truth of the conclusion.

Deductive reasoning mostly starts from the general premises to the specific conclusion, which can be explained as the following example.

Example:

Premise-1: All the human eats veggies.

Premise-2: Suresh is human.

Conclusion: Suresh eats veggies.

The general process of deductive reasoning is given below:

Theory→Hypothesis→Patterns→Confirmation.

(2) Inductive reasoning

Inductive reasoning is a form of reasoning to arrive at a conclusion using limited sets of facts by the process of generalization. It starts with the series of specific facts or data and reaches to a general statement or conclusion.

Inductive reasoning is a type of propositional logic, which is also known as cause-effect reasoning or bottom-up reasoning.

In inductive reasoning, we use historical data or various premises to generate a generic rule, for which premises support the conclusion.

In inductive reasoning, premises provide probable supports to the conclusion, so the truth of premises does not guarantee the truth of the conclusion.

Example:

Premise: All of the pigeons we have seen in the zoo are white.

Conclusion: Therefore, we can expect all the pigeons to be white.

The general process of inductive reasoning is given below:

Observation→Patterns→Hypothesis→Theory.

(3) Abductive reasoning

Abductive reasoning is a form of logical reasoning which starts with single or multiple observations. Then it seeks to find the most likely explanation or conclusion for the observation.

Abductive reasoning is an extension of deductive reasoning, but in abductive reasoning, the premises do not guarantee the conclusion.

Example:

Implication: Cricket ground is wet if it is raining.

Premise: Cricket ground is wet.

Conclusion: It is raining.

(4) Common sense reasoning

Common sense reasoning is an informal form of reasoning, which can be gained through experiences.

Common sense reasoning simulates the human ability to make presumptions about events which occurs on every day.

It relies on good judgment rather than exact logic and operates on heuristic knowledge and heuristic rules.

Example:

1) One person can be at one place at a time.

2) If I put my hand in a fire, then it will burn.

The above two statements are the examples of common sense reasoning which a human mind can easily understand and assume.

(5) Monotonic reasoning

In monotonic reasoning, once the conclusion is taken, then it will remain the same even if we add some other information to existing information in our knowledge base. In monotonic reasoning, adding knowledge does not decrease the set of propositions that can be derived.

To solve monotonic problems, we can derive the valid conclusion from the available facts only, and it will not be affected by new facts.

Monotonic reasoning is not useful for the real-time systems. In real time facts get changed, so we cannot use monotonic reasoning.

Monotonic reasoning is used in conventional reasoning systems and a logic-based system is monotonic.

Any theorem proving is an example of monotonic reasoning.

Example:

Earth revolves around the sun.

It is a fact and it cannot be changed even if we add another sentence in knowledge base like, "The moon revolves around the earth" or "Earth is not round," etc.

The advantages of monotonic reasoning are as follows.

In monotonic reasoning, each old proof will always remain valid.

If we deduce some facts from available facts, then they will remain valid for always.

The disadvantages of monotonic reasoning are as follows.

We cannot represent the real world scenarios using monotonic reasoning.

Since we can only derive conclusions from the old proofs, new knowledge from the real world cannot be added.

(6) Non-monotonic reasoning

Logic will be said as non-monotonic if some conclusions can be invalidated by adding more knowledge into our knowledge base.

In non-monotonic reasoning, some conclusions may be invalidated if we add some more information to our knowledge base.

Non-monotonic reasoning deals with incomplete and uncertain models.

"Human perceptions for various things in daily life", is a general example of non-monotonic reasoning.

Example: Let's suppose the knowledge base contains the following knowledge.

Birds can fly.

Penguins cannot fly.

Pitty is a bird.

So from the above sentences, we can conclude that Pitty can fly.

However, if we add another sentence into knowledge base "Pitty is a penguin", which concludes "Pitty cannot fly", so it invalidates the above conclusion.

The advantages of non-monotonic reasoning are as follows.

For real-world systems such as robot navigation, we can use non-monotonic reasoning.

In non-monotonic reasoning, we can choose probabilistic facts or can make assumptions.

The disadvantages of non-monotonic reasoning are as follows.

In non-monotonic reasoning, the old facts may be invalidated by adding new sentences.

It cannot be used for theorem proving.

✍ New Words

scheme	[ski:m]	n.方案；计划
mental	['mentl]	adj.智力的
infer	[ɪn'fɜ:]	v.推断，推理
rationally	['ræʃnəli]	adv.讲道理地，理性地
valid	['vælɪd]	adj.有效的；正当的
essential	[ɪ'senʃl]	adj.基本的；必不可少的；根本的 n.必需品；基本知识
deductive	[dɪ'dʌktɪv]	adj.推论的，演绎的
inductive	[ɪn'dʌktɪv]	adj.归纳的；归纳法的
abductive	[æb'dʌktɪv]	adj.溯因的；诱导的
monotonic	[ˌmɒnə'tɒnɪk]	adj.单调的，无变化的
argument	['ɑ:gjumənt]	n.论据；讨论，辩论；争论
premise	['premɪs]	n.前提
contradictory	[ˌkɒntrə'dɪktəri]	adj.对立的
hypothesis	[haɪ'pɒθəsɪs]	n.假设，假说
confirmation	[ˌkɒnfə'meɪʃn]	n.证实；确认
generalization	[ˌdʒenrəlaɪ'zeɪʃn]	n.归纳；一般化；普通化
observation	[ˌɒbzə'veɪʃn]	n.观察；监视；评论
explanation	[ˌeksplə'neɪʃn]	n.解释，说明；理由
extension	[ɪk'stenʃn]	n.延伸；扩大
informal	[ɪn'fɔ:ml]	adj.非正式的
presumption	[prɪ'zʌmpʃn]	n.推测，设想
preposition	[ˌprepə'zɪʃn]	n.介词；前置词
proof	[pru:f]	n.证据；证明；检验
represent	[ˌreprɪ'zent]	v.代表；相当于；描绘
invalidate	[ɪn'vælɪdeɪt]	vt.使无效；使作废；证明……错误
incomplete	[ˌɪnkəm'pli:t]	adj.不完备的；不完全的
uncertain	[ʌn'sɜ:tn]	adj.不确定的；多变的
navigation	[ˌnævɪ'geɪʃn]	n.导航
probabilistic	[ˌprɒbəbə'lɪstɪk]	adj.基于概率的；盖然论的
assumption	[ə'sʌmpʃn]	n.假定；承担；获得
sentence	['sentəns]	n.判断

🖎 Phrases

be divided into	划分为
deductive reasoning	演绎推理
inductive reasoning	归纳推理
abductive reasoning	溯因推理
common sense reasoning	常识推理
monotonic reasoning	单调推理
non-monotonic reasoning	非单调推理
propositional logic	命题逻辑
top-down reasoning	自上而下推理
be explained as	被解释为
cause-effect reasoning	因果关系推理
bottom-up reasoning	自下而上推理
generic rule	通用规则
heuristic rule	启发式规则
deal with	处理

Reference Translation

人工智能中的推理

我们已经学习了人工智能中知识表示的各种方式。现在，我们来讨论基于这种知识使用不同逻辑方案进行推理的各种方法。

1. 推理

推理是从现有的知识、事实和信念中得出逻辑结论并做出预测的心理过程。或者我们可以说，"推理是从现有数据中推断事实的一种方式"，它是理性思考以找到有效结论的一般过程。

在人工智能中，推理是必不可少的，因此机器也可以像人的大脑一样理性地思考，并且可以像人一样做事情。

2. 推理类型

在人工智能中，推理可分为以下几类：演绎推理、归纳推理、溯因推理、常识推理、单调推理、非单调推理。

注意：归纳推理和演绎推理是命题逻辑的形式。

（1）演绎推理

演绎推理是从逻辑相关的已知信息中推导出新信息。这是有效推理的形式，这意味着当前提为真时，论据结论必须为真。

演绎推理是人工智能中的一种命题逻辑，它需要各种规则和事实。它有时被称为自上而下的推理，它与归纳推理是相反的。

在演绎推理中，前提的真实性保证了结论的真实性。

演绎推理主要从一般前提到具体结论，可以通过以下示例进行解释。

例子：

前提 1：人类都吃蔬菜。

前提 2：苏雷什是人。

结论：苏雷什吃蔬菜。

演绎推理的一般过程如下：

理论→假设→模式→确认

（2）归纳推理

归纳推理是一种通过归纳过程使用有限的事实集得出结论的推理形式。它从一系列特定的事实或数据开始，直至得出一般性陈述或结论。

归纳推理是命题逻辑的一种，也称为因果推理或自下而上的推理。

在归纳推理中，我们使用历史数据或各种前提来生成通用规则，前提支持结论。

在归纳推理中，前提为结论提供了可能的支持，因此前提的真实性不能保证结论的真实性。

例子：

前提：我们在动物园里看到的所有鸽子都是白色的。

结论：因此，我们可以预期所有鸽子都是白色的。

归纳推理的一般过程如下：

观察→模式→假设→理论

（3）溯因推理

溯因推理是一种逻辑推理的形式，它始于单次或多次观察，然后试图为观察找到最可能的解释或结论。

溯因推理是演绎推理的扩展，但是在溯因推理中，前提并不能保证结论。

例子：

含义：如果下雨，板球场是湿的。

前提：板球场是湿的。

结论：下雨了。

（4）常识推理

常识推理是一种非正式的推理形式，可以通过经验获得。

常识推理模拟了人类对每天发生的事件做出假设的能力。

它依赖于良好的判断力而不是确切的逻辑，并且依据启发式知识和启发式规则来运作。

例子：

1）一个人一次只能在一个地方。

2）如果我将手放在火中，则会燃烧。

以上两个陈述是人脑可以轻松理解和假定的常识推理示例。

（5）单调推理

在单调推理中，一旦得出结论，即使我们将一些其他信息添加到知识库的现有信息中，结论也将保持不变。在单调推理中，添加知识不会减少可以派生的命题集。

为了解决单调问题，我们只能从现有事实中得出有效结论，而不会受到新事实的影响。

单调推理对实时系统没有作用。实时事实会发生变化，因此我们不能使用单调推理。

在常规推理系统中使用单调推理，而基于逻辑的系统是单调的。

任何定理证明都是单调推理的一个例子。

例子：

地球围绕太阳旋转。

这是一个事实，即使我们在知识库中添加诸如"月亮绕地球旋转"或"地球不圆"之类的句子，也无法更改。

单调推理的优点如下。

在单调推理中，每个旧的证据将始终保持有效。

如果我们从现有事实中推断出一些事实，那么这些事实将永远有效。

单调推理的缺点如下。

我们不能使用单调推理来描述现实世界场景。

由于我们只能从旧的证据中得出结论，因此无法添加来自现实世界的新知识。

（6）非单调推理

如果通过在我们的知识库中添加更多知识可以使某些结论无效，那么逻辑将被称为非单调的。

在非单调推理中，如果我们在知识库中添加更多信息，则某些结论可能无效。

非单调推理处理的是不完整和不确定的模型。

"人们对日常生活中各种事物的看法"是非单调推理的一般示例。

示例：假设知识库包含以下知识：

鸟儿会飞。

企鹅不会飞。

皮蒂是一只鸟儿。

因此，从以上句子中，我们可以得出结论，皮蒂会飞。

但是，如果我们在知识库中添加另一个句子"皮蒂是企鹅"，得出的结论是"皮蒂不会飞"，那么上述结论将无效。

非单调推理的优点如下。

对于机器人导航等现实系统，我们可以使用非单调推理。

在非单调推理中，我们可以选择概率事实或做出假设。

非单调推理的缺点如下。

在非单调推理中，通过添加新句子可以使旧的事实无效。

它不能用于定理证明。

Text B

Fuzzy Logic System

扫码听课文

1. Introduction to fuzzy logic system

Fuzzy logic is a computing approach based on "Degree of Truth" and is not limited to boolean "true or false". The term "fuzzy" means something vague or not very clear. The fuzzy logic system is applied to scenarios where it is difficult to categorize states as a binary "True or False". Fuzzy logic can incorporate intermediate values like partially true and partially false. It can be implemented across a wide range of devices ranging from small micro-controller to large IT systems. It tries to mimic human-like decision making, which can incorporate all values in between True and False.

2. Architecture of fuzzy logic system

The fuzzy logic system has four major components, which are explained with the help of the architecture diagram (Figure 3-1) below.

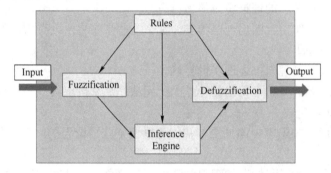

Figure 3-1　architecture diagram

- Rules: Rule base consists of a large set of rules programmed and fed by experts that govern the decision making in the fuzzy system. The rules are sets of "if-then" statements that decide the event occurrence based on condition.
- Fuzzification: Fuzzification converts raw inputs measured from sensors into fuzzy sets. These converted inputs are passed on to the control system for further processing.
- Inference engine: It helps in mapping rules to the input dataset and thereby decide which rules are to be applied for a given input. It does so by calculating the % match of the rules for the given input.
- Defuzzification: It is the opposite of fuzzification. Here fuzzy sets are converted into crisp inputs. These crisp inputs are the output of the fuzzy logic system.

3. Membership function

The membership function defines how input in the fuzzy system is mapped to values between 0 and 1. Input is usually termed as Universe (U) as it can contain any value. The membership function is defined as

$$\mu A{:}X \rightarrow [0,1]$$

The Triangular membership function is the most commonly used membership function. Other membership function includes Trapezoidal, Gaussian and Singleton.

4. Why and when to use fuzzy logic

Fuzzy logic is especially useful when you want to mimic human-like thinking in a control system. It is designed to deal with uncertainties and is proficient in finding out inference from the conclusion.

5. Algorithm of fuzzy logic system

- It defines all the variables and terms which will be acting as input in the fuzzy system.
- It creates membership function for the system (as defined above).
- It creates rule base which will be mapped to each input.
- It converts normal input into fuzzy input and feed it to the membership function.
- It evaluates the result from the membership function.
- It combines all the result obtained from the individual ruleset.
- It converts the output fuzzy set into crisp input (defuzzification).

6. Application of fuzzy logic system

Fuzzy logic has been adopted across all major industries, but automotive remains the major adopters. Some of its applications are listed below.

- Nissan uses fuzzy logic to control the braking system in case of a hazard. Fuzzy logic uses inputs like speed, acceleration and momentum to decide on brakes intensity.
- Nissan also uses fuzzy logic to control the fuel injection quantity and ignition based on inputs like engine RPM, temperature and load capacity.
- It is used in satellites and aircraft for altitude control.
- Mitsubishi uses fuzzy logic to make elevator management more efficient by taking passenger traffic as input.
- Nippon Steel uses fuzzy logic to decide the proportion in which different cement types should be mixed to make more durable cement.
- Fuzzy logic finds its application in the chemical industry for managing the different processes like PH control, drying process and distillation process.
- Fuzzy logic can be combined with artificial neural network to mimic how a human brain works. Fuzzy logic aggregates data and transforms into more meaningful information which is used as fuzzy sets.

7. Advantages and disadvantages of fuzzy logic system

The following are five advantages of the fuzzy logic system.

- Fuzzy logic can work with any kind of input even if it is unstructured, distorted, imprecise or contain noise.
- Fuzzy logic construction is very easy to read and comprehend as it closely mimics the way human-mind makes the decision.
- Fuzzy Logic's nuances involve the use of key maths concepts like set theory and probability, which makes it apt to solve all kinds of day-to-day challenges that humanity faces.
- Fuzzy logic can provide efficient solutions to a very complex problem across different industries.
- Fuzzy logic system needs a very little amount of data to prepare a robust model. Therefore, it needs only a limited amount of memory for its execution.

The following are the top four disadvantages of the fuzzy logic system.

- There is no standard way to solve a problem through fuzzy logic. Therefore, different experts may have a different solution to a problem which leads to ambiguity.

- As fuzzy logic system works with precise and imprecise data, its accuracy can be compromised at times.
- Fuzzy logic system cannot learn from its past mistakes or failures as it doesn't have self-learning ability like machine learning and neural network.
- Due to the lack of standardization, there is no one fixed way to find rules and membership functions for the given problem. Therefore, at times it becomes difficult to find exact rules and membership functions for some problems.

8. Conclusion

Fuzzy logic provides an alternative way to approach real-world problems in the computing world. It can be easily applied to different applications and control system, which can reap long term benefits. Given its ability to work well with "Degree of Truth", it opens many doors to modern computing. However, it is not the panacea to all the problems as it has severe limitations when it comes to accuracy and its inability to learn from its failure as in the case of machine learning.

New Words

fuzzy	['fʌzi]	adj.模糊的
degree	[dɪ'griː]	n.级别；程度
boolean	['buːliən]	adj.布尔的
vague	[veɪg]	adj.模糊的；（思想上）不清楚的；（表达或感知）含糊的
incorporate	[ɪn'kɔːpəreɪt]	vt.包含；使混合；使具体化 vi.包含；吸收；合并；混合
partially	['pɑːʃəli]	adv.部分地
micro-controller	[ˌmɪkrəʊ kən'trəʊlə]	n.微控制器
fuzzification	[fʌzɪfɪ'keɪʃn]	n.模糊性
convert	[kən'vɜːt]	v.转换，转变；改造
control	[kən'trəʊl]	v.控制
map	[mæp]	v.映射
dataset	['deɪtəset]	n.数据集
match	[mætʃ]	v.使相配，使相称
defuzzification	[dɪfʌzɪfɪ'keɪʃn]	n.逆模糊化，去模糊化
crisp	[krɪsp]	adj.清晰的；洁净的；挺括的
triangular	[traɪ'æŋgjələ]	adj.三角形的
trapezoidal	[træpɪ'zɔɪdəl]	adj.梯形的
singleton	['sɪŋgltən]	n.单独的人（或物体）
acceptable	[ək'septəbl]	adj.可接受的

proficient	[prə'fɪʃnt]	adj.精通的，熟练的
evaluate	[ɪ'væljueɪt]	v.估计
hazard	['hæzəd]	vt.冒险；使遭受危险
		n.危险；冒险的事
momentum	[mə'mentəm]	n.动量
intensity	[ɪn'tensəti]	n.强度；烈度
ignition	[ɪg'nɪʃn]	n.（汽油发动机的）点火装置
elevator	['elɪveɪtə]	n.电梯
durable	['djʊərəbl]	adj.耐用的；持久的
distillation	[dɪstɪ'leɪʃn]	n.蒸馏（过程）；蒸馏物
aggregate	['ægrɪgət]	n.总数，合计
		adj.总数的，总计的
	['ægrɪgeɪt]	v.总计；汇集
distort	[dɪ'stɔ:t]	v.歪曲，曲解；（使）变形，失真
imprecise	[ˌɪmprɪ'saɪs]	adj.不精确的，不确定的
noise	[nɔɪz]	n.噪声；干扰信息
comprehend	[ˌkɒmprɪ'hend]	v.理解，领悟
humanity	[hju:'mænəti]	n.人类；人性；人道
robust	[rəʊ'bʌst]	adj.健壮的，强健的；结实的
standard	['stændəd]	n.标准
		adj.标准的
compromise	['kɒmprəmaɪz]	n.折中；妥协方案；达成协议
		v.妥协；使陷入危险，损害；泄露
exact	[ɪg'zækt]	adj.精确的；确切的
reap	[ri:p]	v.获得；得到（报偿）
panacea	[ˌpænə'si:ə]	n.灵丹妙药
inability	[ˌɪnə'bɪləti]	n.无能，无力

🕊 Phrases

fuzzy logic system	模糊逻辑系统
be applied to	适用于，应用于
intermediate value	中间值
architecture diagram	体系结构图
raw input	原始输入
fuzzy set	模糊集合
inference engine	推理机，推理工具
membership function	隶属函数

be defined as	被定义为
braking system	制动系统
fuel injection quantity	燃油喷射量
load capacity	负载容量；载重量
altitude control	高度控制
set theory	集合论

✍ Abbreviations

IT (Information Technology)	信息技术
RPM (Revolutions Per Minute)	转数/分

Reference Translation

模糊逻辑系统

1．模糊逻辑系统简介

　　模糊逻辑是一种基于"真实度"的计算方法，不限于布尔值"真或假"。"模糊"一词的意思是含糊不清或不太清楚。模糊逻辑系统应用于难以将状态归类为二元的"真或假"的场景。模糊逻辑可以包含中间值，如部分为真和部分为假。它可以在从小型微控制器到大型 IT 系统的各种设备中实现。它试图模仿类似于人类的决策，可以包含介于"真"与"假"的所有值。

2．模糊逻辑系统架构

　　模糊逻辑系统具有四个主要组成部分，如下面的体系结构图（见图 3-1）所示。

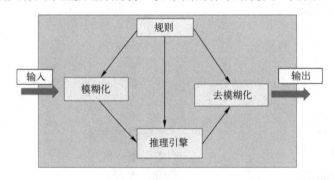

图 3-1　体系结构图

- 规则：规则库由大量规则组成，这些规则由专家编程并输入，用于管理模糊系统中的决策。规则是一组"if-then"语句，它们根据条件决定事件的发生。
- 模糊化：模糊化将从传感器测量的原始输入转换为模糊集。这些转换后的输入将传递到控制系统以进行进一步处理。
- 推理引擎：它有助于将规则映射到输入数据集，从而确定应用于给定输入的规则。它通过计算与给定输入匹配的规则百分比来实现。

- 去模糊化：它与模糊化相反。在这里，模糊集被转换为清晰的输入。这些清晰的输入是模糊逻辑系统的输出。

3. 隶属函数

隶属函数定义如何将模糊系统的输入映射到 0 到 1 之间的值。输入通常可以称为 Universe(U)，因为它可以包含任何值。隶属函数定义为

$$\mu A:\ X \rightarrow [0,1]$$

三角隶属函数是最常用的隶属函数。其他隶属函数包括梯形隶属函数、高斯隶属函数和单例隶属函数。

4. 为什么以及何时使用模糊逻辑

当你想在控制系统中模仿人类思维时，模糊逻辑特别有用。它旨在处理不确定性，并且擅长从结论中找出推论。

5. 模糊逻辑系统算法

- 它定义了作为模糊系统输入的所有变量和项。
- 它为系统创建隶属函数（如上定义）。
- 它创建映射到每个输入的规则库。
- 它将常规输入转换为模糊输入，然后将其输入到隶属函数。
- 它评估隶属函数的结果。
- 它合并从单个规则集获得的所有结果。
- 它将输出模糊集转换为清晰的输入（去模糊化）。

6. 模糊逻辑系统的应用

模糊逻辑已在所有主要行业中采用，但汽车行业仍是主要采用者。下面列出了一些应用。

- 日产使用模糊逻辑来控制制动系统，以防发生危险。模糊逻辑使用速度、加速度和动量等输入来确定制动强度。
- 日产还使用模糊逻辑，根据发动机转速、温度和负载能力等输入来控制燃油喷射量和点火。
- 在卫星和飞机上用于高度控制。
- 三菱使用模糊逻辑，以客流量为输入，使电梯管理更加高效。
- Nippon steel 使用模糊逻辑来决定不同类型水泥的混合比例，以制造更耐用的水泥。
- 模糊逻辑在化学工业中得到了应用，可用于管理 pH 控制、干燥过程和蒸馏过程等不同过程。
- 模糊逻辑可以与人工神经网络（ANN）结合使用，以模拟人脑的工作方式。模糊逻辑将数据聚合并转换为更有意义的信息，这些信息将用作模糊集。

7. 模糊逻辑系统的优缺点

以下是模糊逻辑系统的五个优点。

- 模糊逻辑可以用于任何类型的输入，即使它非结构化、失真、不精确或包含噪声。
- 模糊逻辑的构造非常易于阅读和理解，因为它模仿了人类思维的决策方式。
- 模糊逻辑的细微差别涉及诸如集合论和概率之类的关键数学概念的使用，这使其易于解决人类面临的各种日常挑战。

- 模糊逻辑可以为不同行业的非常复杂的问题提供有效的解决方案。
- 模糊逻辑系统只需要很少量的数据就可以构建健壮的模型。因此，它只需要有限数量的内存即可执行。

以下是模糊逻辑系统的四个主要缺点。

- 用模糊逻辑解决问题没有标准方法。因此，对于导致歧义的问题，不同的专家可能有不同的解决方案。
- 由于模糊逻辑系统可以处理精确和不精确的数据，因此有时可能会降低其准确性。
- 模糊逻辑系统不具备机器学习和神经网络等自主学习能力，因此无法从过去的错误或失败中学习。
- 由于缺乏标准化，因此没有一种固定的方法可以找到给定问题的规则和隶属函数。所以，有时很难找到某些问题的确切规则和隶属函数。

8. 结论

模糊逻辑为解决计算世界中的实际问题提供了另一种方法。它可以很容易地运用到不同的应用和控制系统中，从而获得长期的效益。由于能够很好地处理"真实度"，它为现代计算打开了许多大门。然而，它并不是解决所有问题的灵丹妙药，因为它在准确性方面存在严重的局限性，而且无法像机器学习那样从失败中学习。

Exercises

[Ex. 1] Answer the following questions according to Text A.

1. What is reasoning?

2. What are the categories of reasoning in artificial intelligence?

3. What is deductive reasoning?

4. What is inductive reasoning?

5. What is deductive reasoning sometimes referred to as? What is inductive reasoning also known as?

6. What is abductive reasoning?

7. What is common sense reasoning? What does it rely and operate on?

8. What can we do to solve monotonic problems?

9. Is monotonic reasoning useful for the real-time systems? Where is it used?

10. What are the advantages of non-monotonic reasoning?

[Ex. 2] Answer the following questions according to Text B.

1. What is fuzzy logic?

2. How many major components does the fuzzy logic system have? What are they?

3. What does rule base consist of?

4. What does inference engine do? And how?

5. What does the membership function define?

6. When is fuzzy logic especially useful? What is it designed to do?

7. What does algorithm of fuzzy logic system define?

8. What does Nippon Steel use fuzzy logic to do?

9. What is the first advantage of the fuzzy logic system mentioned in the passage?

10. What is the last disadvantage of the fuzzy logic system mentioned in the passage?

[Ex. 3] Translate the following terms or phrases from English into Chinese.

1. abductive   1. _____

2. confirmation   2. _____

3. deductive   3. _____

4. inductive   4. _____

5. infer   5. _____

6. abductive reasoning   6. _____

7. bottom-up reasoning   7. _____

8. common sense reasoning   8. _____

9. non-monotonic reasoning   9. _____

10. architecture diagram   10. _____

[Ex. 4] Translate the following terms or phrases from Chinese into English.

1. *n.*前提   1. _____

2. *n.*推测，设想   2. _____

3. *adj.*基于概率的；盖然论的   3. _____

4. *n.*判断   4. _____

5. *adj.*布尔的   5. _____

6. 模糊集合   6. _____

7. 推理机，推理工具   7. _____

8. 集合论   8. _____

9. 模糊逻辑系统   9. _____

10. 命题逻辑   10. _____

[Ex. 5] Translate the following passage into Chinese.

Inference Engines

Inference engines are a component of an artificial intelligence system. They apply logical rules to a knowledge graph (or knowledge base) to surface new facts and relationships. Implementation of inference engines can proceed via induction or deduction. The process of inferring relationships between entities utilizing machine learning, machine vision, and natural language processing have exponentially increased the scale and value of knowledge graphs and relational databases in recent years.

Historically inference engines surfaced as components within expert systems, systems meant to emulate the problem solving ability of a human expert within a given domain.

Two methods employed within many inference engines to infer new knowledge include what are known as the backward chaining and forward chaining reasoning methods.

- Backward chaining reasoning methods begin with a list of hypotheses and work backwards to see if data, once plugged into rules, support these hypotheses. In short, backward chaining highlights what facts must be true to support a hypothesis.
- Forward chaining reasoning methods start with available data and utilize rules to infer new data. In short, forward chaining starts with known facts and uses them to create new facts.

Both backward and forward chaining reasoning progress according to the modus ponens form of deductive reasoning. In other words, if X is true, then Y is true. X is true, and therefore Y must be true.

In many contemporary AI applications, both backward and forward chaining are applied in what is referred to as opportunistic reasoning.

At its simplest form, the actions performed by inference engines tend to progress through three stages: matching rules, selecting rules and executing rules.

- Matching rules is an action in which an inference engine finds all rules triggered by the contents of a knowledge base.
- Selecting rules is an action which discerns which order rules should be applied in (this will differ for forward or backward chaining, or by other machine learning inputs).
- Executing rules is to apply rules to existing knowledge through forward or backward chaining.

Once executing rules is completed, matching rules is re-started until there are no more opportunities for either forward or backward chaining deductions.

Unit 4

Text A

Informed Search Algorithms

1. Introduction to informed search algorithms

Informed search requires domain knowledge and the details provided in problem definition to search/traverse the search space and reach the goal node, and is therefore considered more efficient.

Informed search algorithm, in short, contains an array of knowledge, like distance to the goal, path cost, steps to reach the goal node, among others, which helps it reach the solution more efficiently, without compromising time and cost of the search. However, to achieve this it uses the heuristic function, which is based on evaluation applied to states in the search tree.

For example, when searching on Google Maps you give the search algorithm information like a place you plan to visit from your current location for it to accurately navigate the distance, the time taken and real-time traffic updates on that particular route. This is all driven by complex informed search algorithms powering Google Maps search functionality.

2. Types of informed search algorithms

Before getting started with different types of informed search algorithms, it's important to understand some basic concepts like search space which refers to space or the database in which the search is to be performed, the initial state or start state from where the search begins and the goal state which is the result of the search like our destination in the earlier example of Google Maps and goal test to check whether the current state is the destination or goal state. Path cost is a numerical term assigned to measure the numeric cost of the path taken to achieve the goal. Heuristic function, which is a function used to measure how close our current state is to the goal state, uses heuristic properties to find out the best possible path with respect to path cost to achieve the goal state.

In informed search algorithms, we have information on the goal state which narrows down our results precisely. There may be many possible ways to get to the goal state, but we need to get the best possible outcome or path for our search; this is where informed search shines.

(1) Pure heuristic search

Pure heuristic search is the simplest form of heuristic search algorithms. It expands nodes based on their heuristic value h(n). It maintains two lists, OPEN and CLOSED list. In the CLOSED list, it places those nodes which have already expanded and in the OPEN list, it places nodes which have yet not been expanded.

On each iteration, each node n with the lowest heuristic value is expanded and generates all its successors and n is placed to the CLOSED list. The algorithm continues until a goal state is found.

(2) Best first search algorithm (greedy search)

Best first search algorithm always selects the path which appears best at that moment. It is the combination of depth first search and breadth first search algorithms. It uses the heuristic function and search. Best first search allows us to take the advantages of both algorithms. With the help of best first search, at each step, we can choose the most promising node. In the best first search algorithm, we expand the node which is closest to the goal node and the closest cost is estimated by heuristic function.

Best first algorithm is implemented by the priority queue.

Best first search algorithm steps are as follows.

1) Place the starting node into the OPEN list.

2) If the OPEN list is empty, stop and return failure.

3) Remove the node n, from the OPEN list which has the lowest value of h(n), and places it in the CLOSED list.

4) Expand the node n, and generate the successors of node n.

5) Check each successor of node n, and find whether any node is a goal node or not. If any successor node is goal node, then return success and terminate the search, else proceed to 6).

6) For each successor node, algorithm checks for evaluation function f(n), and then check if the node has been in either OPEN or CLOSED list. If the node has not been in both list, then add it to the OPEN list.

7) Return to 2).

The advantages of best first search algorithm are as follows.

- Best first search can switch between BFS and DFS by gaining the advantages of both the algorithms.
- This algorithm is more efficient than BFS and DFS algorithms.

The disadvantages of best first search algorithm are as follows.

- It can behave as an unguided depth first search in the worst case scenario.
- It can get stuck in a loop as DFS.
- This algorithm is not optimal.

(3) A* search

A* search is the most commonly known form of best first search. It uses heuristic function h(n), and cost to reach the node n from the start state g(n). It has combined features of UCS (Uniform Cost Search see Chapter Text B) and greedy best first search, by which it solve the problem efficiently. A* search algorithm finds the shortest path through the search space using the heuristic function. This search algorithm expands less search tree and provides optimal result faster. A* algorithm is similar to UCS except that it uses evaluation function g(n)+h(n) instead of g(n).

In A* search algorithm, we use search heuristic and calculate the cost to reach the node. Hence we can combine both costs as following, and this sum is called as a fitness number.

$$f(n)=g(n)+h(n)$$

f(n): Estimated cost of the cheapest solution.

g(n): Cost to reach node n from start state.

h(n): Cost to reach node n from goal state.

At each point in the search space, only those nodes are expanded which have the lowest value of f(n), and the algorithm terminates when the goal node is found.

Algorithm of A* search steps are as follows.

1) Place the starting node in the OPEN list.

2) Check if the OPEN list is empty or not, if the list is empty then return failure and stops.

3) Select the node from the OPEN list which has the smallest value of evaluation function (g+h), if node n is goal node then return success and stop, otherwise.

4) Expand node n and generate all of its successors, and put n into the CLOSED list. For each successor n', check whether n' is already in the OPEN or CLOSED list, if not then compute evaluation function for n' and place into OPEN list.

5) Else if node n' is already in OPEN and CLOSED list, then it should be attached to the back pointer which reflects the lowest g(n') value.

6) Return to 2).

The advantages of A* search are as follows.

- A* search algorithm is the best algorithm compared with search algorithms.
- A* search algorithm is optimal and complete.
- This algorithm can solve very complex problems.

The disadvantages of A* search are as follows.

- It does not always produce the shortest path as it is mostly based on heuristics and approximation.
- A* search algorithm has some complexity issues.
- The main drawback of A* search algorithm is memory requirement as it keeps all generated nodes in the memory. So it is not practical for various large-scale problems.

3. Conclusion

"Search" is a very important aspect of getting the desired result out of a very large, complex and ever-growing database of any organization or institution around the world. It is imperative to have a working knowledge of different types of search in the study of data science in general and database management systems in particular. With the advent of AI and ML, there are astounding opportunities to develop and create new ways to search databases and provide users with answers that help them solve their problems.

✑ New Words

algorithm	['ælgərɪðəm]	n.算法
definition	[ˌdefɪ'nɪʃn]	n.定义
search	[sɜ:tʃ]	v.&n.搜索
traverse	[trə'vɜ:s]	v.遍历

node	[nəʊd]	*n.*节点
evaluation	[ɪˌvælju'eɪʃn]	*n.*评估，评价
state	[steɪt]	*n.*状态
particular	[pə'tɪkjələ]	*adj.*特定的，特殊的
route	[ru:t]	*n.*路
concept	['kɒnsept]	*n.*概念；观念
perform	[pə'fɔ:m]	*v.*执行；起……作用
measure	['meʒə]	*v.*衡量；测量；量度；估量
precisely	[prɪ'saɪsli]	*adv.*精确地
expand	[ɪk'spænd]	*v.*扩大，扩展
list	[lɪst]	*n.*列表；清单
iteration	[ˌɪtə'reɪʃn]	*n.*迭代；循环
successor	[sək'sesə]	*n.*后继者，继任者
select	[sɪ'lekt]	*v.*选择，（在计算机屏幕上）选定
combination	[ˌkɒmbɪ'neɪʃn]	*n.*结合（体）；联合（体）
promising	['prɒmɪsɪŋ]	*adj.*有希望的；有前途的
priority	[praɪ'ɒrəti]	*n.*优先级，优先权
queue	[kju:]	*n.*队列
empty	['empti]	*adj.*空的
return	[rɪ'tɜ:n]	*v.*返回
failure	['feɪljə]	*n.*失败；故障
terminate	['tɜ:mɪneɪt]	*v.*结束；使终结
check	[tʃek]	*v.*检查；查看；核实
switch	[swɪtʃ]	*v.*切换；改变，转变
unguided	['ʌn'gaɪdɪd]	*n.*无向导的；不能控制的
loop	[lu:p]	*n.*循环
combine	[kəm'baɪn]	*v.*使结合，使合并
estimate	['estɪmət]	*n.*估计，估算；评价
	['estɪmeɪt]	*vt.*估计，估算；评价
reflect	[rɪ'flekt]	*v.*反映；反射
complete	[kəm'pli:t]	*adj.*完全的，完整的
approximation	[əˌprɒksɪ'meɪʃn]	*n.*近似法，近似值
aspect	['æspekt]	*n.*方面；样子
institution	[ˌɪnstɪ'tju:ʃn]	*n.*机构
astound	[ə'staʊnd]	*vt.*使震惊，使大吃一惊
opportunity	[ˌɒpə'tju:nəti]	*n.*机会；时机

🖎 Phrases

informed search	知情搜索
search space	搜索空间
goal node	目标节点
an array of	一排，一群，一批
path cost	路径成本
heuristic function	启发函数
search tree	搜索树
current location	当前位置
initial state	初始状态，初态
goal state	目标状态
current state	当前状态
narrow down	变窄，减少，缩小
pure heuristic search	纯启发式搜索
heuristic value	启发值
best first search	最佳优先搜索
greedy search	贪婪搜索
worst case scenario	最坏情况
greedy best-first search	贪婪最佳优先搜索
memory requirement	内存需求（量）
large-scale problem	大规模问题，大型问题
data science	数据科学
database management system	数据库管理系统

🖎 Abbreviations

DFS (Depth First Search)	深度优先搜索
BFS (Breadth First Search)	宽度优先搜索
UCS (Uniform Cost Search)	统一成本搜索

Reference Translation

知情搜索算法

1. 知情搜索算法简介

　　知情搜索算法需要领域的知识和问题定义中提供的详细信息，才能搜索/遍历搜索空间并到达目标节点，因此被认为更有效。

简而言之，知情搜索算法包含一系列知识，如到目标的距离、路径成本、到达目标节点的步骤等，这些知识有助于在不影响搜索时间和成本的情况下更有效地实现解决方案。但是，为实现此目的，它使用了启发函数，该函数基于应用于搜索树中状态的评估。

例如，在谷歌地图上搜索时，你会给搜索算法提供信息，如从当前位置计划到达的地点，这样它就能准确地导航，告知该路线的距离、所花费的时间和实时路况。这一切都是由支持谷歌地图搜索功能的复杂的知情搜索算法驱动的。

2. 知情搜索算法的类型

在开始使用不同类型的知情搜索算法之前，重要的是要了解一些基本概念，如搜索空间（指要在其中执行搜索的空间或数据库）、搜索开始的初始状态或开始状态以及目标状态（这是搜索的结果），就像我们在前面的谷歌地图和目标测试示例中的目标位置一样，目的是检查当前状态是不是目的或目标状态。路径成本是一个数字术语，用来度量为实现目标而采取的路径的数字成本。启发函数是一种用于度量当前状态与目标状态之间的接近程度的函数，它使用启发属性来找出实现目标状态所需的最佳成本的路径。

在知情搜索算法中，我们具有目标状态的信息，从而可以精确地缩小搜索范围。达到目标状态可能有很多方法，但是我们需要找到最佳结果或搜索路径。这就是知情搜索的亮点。

（1）纯启发式搜索

纯启发式搜索是启发式搜索算法的最简单形式。它根据节点的启发值 h(n)扩展节点。它维护两个列表：OPEN 和 CLOSED 列表。在 CLOSED 列表中，它放置了已经扩展的节点，在 OPEN 列表中，它放置了尚未扩展的节点。

在每次迭代中，具有最低启发值的每个节点 n 都会被扩展并生成其所有后继节点，并且 n 会被放置到 CLOSED 列表中。该算法继续运行，直到找到目标状态为止。

（2）最佳优先搜索算法（贪婪搜索）

最佳优先搜索算法始终会选择此时此刻出现的最佳路径。它是深度优先搜索算法和宽度优先搜索算法的结合。它使用启发函数和搜索。最佳优先搜索使我们能够利用两种算法的优势。借助最佳优先搜索，我们可以在每个步骤中选择最有前途的节点。在最佳的优先搜索算法中，我们扩展最接近目标节点的节点，并通过启发函数估算最接近的成本。

最佳优先算法是由优先级队列实现的。

最佳优先搜索算法的步骤如下。

1）将起始节点放入 OPEN 列表中。

2）如果 OPEN 列表为空，则停止并返回失败。

3）从 OPEN 列表中删除具有最低 h(n)值的节点 n，并将其放在 CLOSED 列表中。

4）展开节点 n，并生成节点 n 的后继节点。

5）检查节点 n 的每个后继节点，并确定是否有某个节点是目标节点。如果有一个后继节点是目标节点，则返回成功并终止搜索，否则继续执行步骤 6）。

6）对于每个后继节点，算法将检查评估函数 f(n)，然后检查该节点是否已在 OPEN 列表或 CLOSED 列表中。如果该节点不在两个列表中，则将其添加到 OPEN 列表中。

7）返回到步骤 2）。

最佳优先搜索算法的优点如下。

● 通过获得两种算法的优势，最佳优先搜索可以在 BFS 和 DFS 之间切换。

● 该算法比 BFS 和 DFS 算法更有效。

最佳优先搜索算法的缺点如下。

● 在最坏的情况下，它可以充当无向导的深度优先搜索。

● 它可能像 DFS 一样会陷入循环。

● 此算法不是最优算法。

（3）A*搜索

A*搜索是最佳优先搜索的最普遍形式。它使用启发函数 h(n) 和从起始状态到达节点 n 的成本函数 g(n)。它结合了统一成本搜索（见本章 Text B）和最佳优先搜索的特点，可以有效地解决问题。A*搜索算法使用启发函数查找通过搜索空间的最短路径。该搜索算法扩展了较少的搜索树，并更快地提供了最佳结果。A*算法与统一成本搜索相似，不同之处在于它使用评估函数 g(n)+ h(n) 代替 g(n)。

在 A*搜索算法中，我们使用启发式搜索并计算到达节点的成本。因此，我们可以将以下两种成本进行合并，该总和称为适应度数。

f(n)= g(n)+ h(n)

f(n)：最便宜的解决方案的估计成本。

g(n)：从开始状态到达节点 n 的成本。

h(n)：从目标状态到达节点 n 的成本。

在搜索空间的每个点上，仅展开那些 f(n) 值最小的节点，并且当找到目标节点时算法终止。

A*搜索算法的步骤如下。

1）将起始节点放置在 OPEN 列表中。

2）检查 OPEN 列表是否为空，如果列表为空，则返回失败并停止。

3）从 OPEN 列表中选择评估函数 g(n)+h(n) 值最小的节点，如果节点 n 是目标节点，则返回成功并停止搜索，否则进行步骤 4）。

4）展开节点 n 并生成其所有后继节点，然后将 n 放入 CLOSED 列表。对于每个后继 n'，检查 n' 是否已在 OPEN 或 CLOSED 列表中，如果不在，则计算 n' 的评估函数并放入 OPEN 列表中。

5）否则，如果节点 n' 已经在 OPEN 和 CLOSED 列表中，则应将其附加到后向指针上，该指针反映出最低的 g(n') 值。

6）返回到步骤 2）。

A*搜索的优点如下。

● 与其他搜索算法相比，A*搜索算法是最佳算法。

● A*搜索算法是最优且完整的。

● 该算法可以解决非常复杂的问题。

A*搜索的缺点如下。

● 它并不总是产生最短的路径，因为它主要基于启发式和近似法。

● A*搜索算法存在一些复杂性问题。

● A*搜索算法的主要缺点是内存需求，因为它会将所有生成的节点都保留在内存中。因此对于各种大规模问题，A*搜索不切实际。

3. 结论

"搜索"是一个非常重要的方面，它可以从全球任何组织或机构的庞大、复杂且不断增长的数据库中获得期望的结果。在研究数据科学（特别是数据库管理系统）时，必须掌握不同类型的搜索工作知识。随着 AI 和 ML 的出现，开发和创建搜索数据库并为用户解决问题提供帮助的新方法的机会多得惊人。

Text B

Uninformed Search Algorithms

扫码听课文

Uninformed search is a class of general-purpose search algorithms which operates in a brute force way. Uninformed search algorithms do not have additional information about state or search space other than how to traverse the tree, so it is also called blind search.

1. Breadth First Search (BFS)

Breadth first search is the most common search strategy for traversing a tree or graph. This algorithm searches breadthwise in a tree or graph, so it is called breadth first search. It starts searching from the root node of the tree and expands all successor node at the current level before moving to nodes of next level.

Breadth first search algorithm is an example of a general-graph search algorithm. It is implemented using FIFO queue data structure.

Advantages: BFS will provide a solution if any solution exists. If there are more than one solutions for a given problem, then BFS will provide the minimal solution which requires the least number of steps.

Disadvantages: It requires lots of memory since each level of the tree must be saved into memory to expand the next level. BFS needs lots of time if the solution is far away from the root node.

Completeness: BFS is complete, which means if the shallowest goal node is at some finite depth, then BFS will find a solution.

Optimality: BFS is optimal if path cost is a non-decreasing function of the depth of the node.

2. Depth First Search (DFS)

Depth first search is a recursive algorithm for traversing a tree or graph data structure. It is called the depth first search because it starts from the root node and follows each path to its greatest depth node before moving to the next path. DFS uses a stack data structure for its implementation. The process of the DFS algorithm is similar to the BFS algorithm.

Note: Backtracking is an algorithm technique for finding all possible solutions using recursion.

Advantages: DFS requires very less memory as it only needs to store a stack of the nodes on the path from root node to the current node. It takes less time to reach to the goal node than BFS algorithm (if it traverses in the right path).

Disadvantages: There is the possibility that many states keep re-occurring, and there is no guarantee of finding the solution. DFS algorithm goes for deep down searching and sometime it may go to the infinite loop.

Completeness: DFS search algorithm is complete within finite state space as it will expand every node within a limited search tree.

Optimality: DFS search algorithm is non-optimal, as it may generate a large number of steps or high cost to reach to the goal node.

3. Depth Limited Search (DLS)

A depth limited search algorithm is similar to depth first search with a predetermined limit. Depth limited search can solve the drawback of the infinite path in the depth first search. In this algorithm, the node at the depth limit will treat as it has no successor nodes further.

Depth limited search can be terminated with two conditions of failure.

Standard failure value: It indicates that problem does not have any solution.

Cutoff failure value: It defines no solution for the problem within a given depth limit.

Advantages: Depth limited search is memory efficient.

Disadvantages: Depth limited search also has a disadvantage of incompleteness. It may not be optimal if the problem has more than one solution.

Completeness: DLS search algorithm is complete if the solution is within the depth limit.

Optimality: Depth limited search can be viewed as a special case of DFS.

4. Uniform Cost Search (UCS)

Uniform cost search is a searching algorithm used for traversing a weighted tree or graph. This algorithm comes into play when a different cost is available for each edge. The primary goal of the uniform cost search is to find a path to the goal node which has the lowest cumulative cost. Uniform cost search expands nodes according to their path costs from the root node. It can be used to solve any graph/tree where the optimal cost is in demand. A uniform cost search algorithm is implemented by the priority queue. It gives maximum priority to the lowest cumulative cost. Uniform cost search is equivalent to BFS algorithm if the path cost of all edges is the same.

Advantages: Uniform cost search is optimal because at every state the path with the least cost is chosen.

Disadvantages: It does not care about the number of steps involved in searching and it is only concerned about path cost, due to which this algorithm may be stuck in an infinite loop.

Completeness: Uniform cost search is complete, such as if there is a solution, UCS will find it.

Optimality: Uniform cost search is always optimal as it only selects a path with the lowest path cost.

5. Iterative Deepening Depth First Search (IDDFS)

The iterative deepening algorithm is a combination of DFS and BFS algorithms. This search algorithm finds out the best depth limit and does it by gradually increasing the limit until a goal is found. This algorithm performs depth first search up to a certain "depth limit", and it keeps increasing the depth limit after each iteration until the goal node is found.

This search algorithm combines the benefits of breadth first search's fast search and depth first search's memory efficiency. It is a useful uninformed search when search space is large, and depth of goal node is unknown.

Advantages: It combines the benefits of BFS and DFS search algorithm in terms of fast search and memory efficiency.

Disadvantages: The main drawback of IDDFS is that it repeats all the work of the previous phase.

Completeness: This algorithm is complete if the branching factor is finite.

Optimality: IDDFS algorithm is optimal if path cost is a non-decreasing function of the depth of the node.

6. Bidirectional Search

Bidirectional search algorithm runs two simultaneous searches, one from initial state called as forward search and the other from goal node called as backward search. It replaces one single search graph with two small subgraphs in which one starts the search from an initial vertex and other starts from goal vertex. The search stops when these two graphs intersect each other. Bidirectional search can use search techniques such as BFS, DFS, DLS, etc.

Advantages: Bidirectional search is fast. Bidirectional search requires less memory.

Disadvantages: Implementation of the bidirectional search tree is difficult. In bidirectional search, one should know the goal state in advance.

Completeness: Bidirectional search is complete if we use BFS in both searches.

Optimality: Bidirectional search is optimal.

✒ New Words

blind	[blaɪnd]	adj.盲目的；不理性的
strategy	['strætədʒi]	n.策略；部署；战略
breadthwise	['bredθwaiz]	adv.横向地
graph	[ɡrɑ:f]	n.图
implement	['ɪmplɪment]	vt.实施，执行
minimal	['mɪnɪml]	adj.最小的，极小的；极少的
solution	[sə'lu:ʃn]	n.解决办法；答案
complexity	[kəm'pleksəti]	n.复杂度，复杂性
frontier	['frʌntɪə]	n.边界；前沿；新领域
completeness	[kəm'pli:tnəs]	n.完备性；完全性；完整性
finite	['faɪnaɪt]	adj.有限的；限定的
optimality	[ɔpti'mæliti]	n.最优性；最佳性
optimal	['ɒptɪməl]	adj.最佳的，最优的
recursive	[rɪ'kɜ:sɪv]	adj.递归的；回归的
backtrack	['bæktræk]	vi.回溯

recursion	[rɪ'kɜːʃn]	n.递归，递归式；递推
memory	['meməri]	n.存储器，内存
store	[stɔː]	v.存储，保存；记忆
infinite	['ɪnfɪnət]	adj.无限的，无穷的
predetermine	[ˌpriːdɪ'tɜːmɪn]	v.预先决定；事先安排
incompleteness	[ˌɪnkəm'pliːtnəs]	n.不完备性，不完整性
cumulative	['kjuːmjələtɪv]	adj.积累的，累计的
concern	[kən'sɜːn]	v.考虑；关心；影响；涉及
worst-case	['wɜːst keɪs]	adj.最坏情况的
gradually	['grædʒuəli]	adv.逐渐地，逐步地
repeat	[rɪ'piːt]	v.重复
simultaneous	[ˌsɪml'teɪniəs]	adj.同时的
replace	[rɪ'pleɪs]	v.替换；以……取代；更新
subgraph	['sʌbɡrɑːf]	n.子图
vertex	['vɜːteks]	n.顶点
intersect	[ˌɪntə'sekt]	v.相交，交叉

✍ Phrases

uninformed search	不知情搜索
brute force	暴力，蛮力
blind search	盲目搜索
root node	根节点
successor node	子节点，后继节点
general-graph search	通用图搜索
data structure	数据结构
be saved into	被保存到
time complexity	时间复杂度
be obtained by	由……得到，由……获得
space complexity	空间复杂度
non-decreasing function	非递减函数
be similar to	类似于，相似于
current node	当前节点
infinite loop	无限循环，无穷循环
be equivalent to	等于，等同于
weighted tree	加权树
according to	根据

be stuck in	困于，停止不前，动弹不得
find out	发现，找出
branching factor	分支因子
bidirectional search	双向搜索
forward search	正向搜索
backward search	反向搜索

✍ Abbreviations

FIFO (First Input First Output)	先进先出
DLS (Depth Limited Search)	限制深度搜索
IDDFS (Iterative Deepening Depth First Search)	迭代加深深度优先搜索

Reference Translation

不知情搜索算法

不知情搜索是一类以暴力方式运行的通用搜索算法。不知情搜索除了遍历树外没有关于状态或搜索空间的其他信息，因此也称为盲搜索。

1．广度优先搜索（BFS）

广度优先搜索是遍历树或图的最常见搜索策略。该算法在树或图中进行广度搜索，因此被称为广度优先搜索。它从树的根节点开始搜索，并在移动到下一级别的节点之前扩展当前级别的所有后续节点。

广度优先搜索算法是通用图搜索算法的示例。它是使用 FIFO 队列数据结构实现的。

优点：如果存在任何解决方案，BFS 将提供解决方案。如果一个给定的问题有多个解决方案，那么 BFS 将提供需要最少步骤的最小解决方案。

缺点：由于树的每个级别都必须保存到内存中才能扩展到下一个级别，因此需要大量内存。如果解决方案离根节点很远，那么 BFS 需要很长时间。

完整性：BFS 是完整的，这意味着如果最浅的目标节点处于某个有限深度，则 BFS 将找到解决方案。

最优性：如果路径成本是节点深度的非递减函数，则 BFS 最优。

2．深度优先搜索（DFS）

深度优先搜索是一种用于遍历树或图数据结构的递归算法。之所以被称为深度优先搜索，是因为它从根节点开始，然后沿着每条路径到达其最大深度节点，然后再移动到下一条路径。DFS 是使用堆栈数据结构实现的。DFS 算法的过程类似于 BFS 算法。

注意：回溯是一种使用递归查找所有可能解的算法技术。

优点：DFS 仅需要存储从根节点到当前节点的路径上的众多节点，因此所需的内存非常少。与 BFS 算法相比，到达目标节点所需的时间更少（如果它遍历正确的路径）。

缺点：有可能许多状态重复出现，并且不能保证找到解决方案。DFS 算法适用于深度搜索，有时可能会进入无限循环。

完整性：DFS 搜索算法在有限的状态空间内是完整的，因为它将扩展有限搜索树中的每个节点。

最优性：DFS 搜索算法不是最佳算法，因为它可能会产生大量步骤，或到达目标节点的成本较高。

3. 深度受限搜索（DLS）

深度受限搜索算法类似于具有预定限制的深度优先搜索。深度受限搜索可以解决深度优先搜索中无限路径的缺点。在该算法中，处于深度限制的节点将被视为没有进一步的后续节点。

深度受限搜索可以以两种失败条件终止。

- 标准失败值：表明问题没有任何解决方案。
- 截止失败值：在给定的深度限制内，该问题没有解决方案。

优点：深度受限搜索提高了存储效率。

缺点：深度受限搜索还具有不完整的缺点。如果问题有多个解决方案，则可能不是最佳选择。

完整性：如果解决方案在深度限制之内，则 DLS 搜索算法是完整的。

最优性：限制深度搜索可以视为 DFS 的一种特殊情况。

4. 统一成本搜索（UCS）

统一成本搜索是用于遍历加权树或图形的搜索算法。当每条边的可用成本不同时，该算法就会发挥作用。统一成本搜索的主要目标是找到一条通往目标节点的路径，该路径具有最低的累积成本。统一成本搜索根据节点从根节点开始的路径成本来扩展节点。它可用于求解需要最佳成本的任何图/树。统一成本搜索算法由优先级队列实现。它为最低的累积成本提供了最高的优先级。如果所有边的路径成本相同，则统一成本搜索等效于 BFS 算法。

优点：统一成本搜索是最佳的，因为在每个状态下都选择了成本最低的路径。

缺点：它不关心搜索中涉及的步骤数，而仅关心路径成本，因此该算法可能陷入无限循环。

完整性：统一成本搜索是完整的，例如，如果有解决方案，UCS 会找到它。

最优性：统一成本搜索始终是最佳选择，因为它只会选择路径成本最低的路径。

5. 迭代加深深度优先搜索（IDDFS）

迭代加深算法是 DFS 和 BFS 算法的组合。该搜索算法找出最佳的深度极限，并逐渐增加极限直到找到目标为止。该算法执行深度优先搜索直到某个"深度限制"，并且在每次迭代之后一直增加深度限制，直到找到目标节点为止。

该搜索算法结合了广度优先搜索的快速搜索和深度优先搜索的存储高效的优点。当搜索空间很大且目标节点的深度未知时，它是有用的不知情搜索。

优点：它结合了 BFS 和 DFS 搜索算法在快速搜索和存储高效的优势。

缺点：它重复了上一阶段的所有工作。

完整性：如果分支因子是有限的，则此算法是完整的。

最优性：如果路径成本是节点深度的非递减函数，则 IDDFS 算法是最佳的。

6. 双向搜索

双向搜索算法同时运行两次搜索，一次从初始状态出发称为正向搜索，另一次从目标节点出发称为反向搜索。双向搜索将一个单独的搜索图替换为两个小的子图，其中一个从初始顶点开始搜索，另一个从目标顶点开始搜索。当这两个图彼此相交时，搜索将停止。双向搜索可以使用 BFS、DFS、DLS 等搜索技术。

优点：双向搜索速度快，且需要更少的内存。

缺点：双向搜索树的实现很困难。在双向搜索中，应该事先知道目标状态。

完整性：如果我们在两个搜索中都使用 BFS，则双向搜索就是完整的。

最优性：双向搜索是最佳的。

Exercises

[Ex. 1] Answer the following questions according to Text A.

1. What does informed search require to search/traverse the search space and reach the goal node?

2. What does informed search algorithm contain in short?

3. What is path cost?

4. What is heuristic function? What does it do?

5. How does pure heuristic search expand nodes? What are the two lists it maintains?

6. What does best first search algorithm always do?

7. What are the advantages of best first search?

8. What is A* search?

9. What are the advantages of A* search algorithm?

10. What is the main drawback of A* search algorithm? Why?

[Ex. 2] Answer the following questions according to Text B.

1. What is uninformed search?

2. Why is uninformed search also called blind search?

3. What is breadth first search? Why is it called so?

4. What are the advantages of BFS?

5. What is depth first search? Why is it called so?

6. What are the disadvantages of DFS?

7. What can depth limited search do? What is the advantage of depth limited search?

8. What is the primary goal of the uniform cost search? Why is it optimal?

9. What is the iterative deepening algorithm? When is it a useful uninformed search?

10. What are the two simultaneous searches bidirectional search algorithm runs? What are its disadvantages?

[Ex. 3] Translate the following terms or phrases from English into Chinese.

1. algorithm   1. _____

2. complete   2. _____

3. evaluation   3. _____

4. iteration   4. _____

5. loop   5. _____

6. best first search   6. _____

7. greedy best first search   7. _____

8. pure heuristic search   8. _____

9. current location   9. _____

10. heuristic function   10. _____

[Ex. 4] Translate the following terms or phrases from Chinese into English.

1. *n.*列表；清单   1. _____

2. *n.*节点   2. _____

3. *n.*优先级，优先权   3. _____

4. *n.*队列   4. _____

5. *v.&n.*搜索   5. _____

6. 初始状态，初态   6. _____

7. 路径成本   7. _____

8. 目标节点   8. _____

9. 加权树   9. _____

10. 反向搜索   10. _____

[Ex. 5] Translate the following passage into Chinese.

Genetic Algorithm

A genetic algorithm is a heuristic search method used in artificial intelligence and computing. It is used for finding optimized solutions to search problems based on the theory of natural selection and evolutionary biology. Genetic algorithms are excellent for searching through large and complex data sets. They are considered capable of finding reasonable solutions to complex issues as they are highly capable of solving unconstrained and constrained optimization issues.

A genetic algorithm makes uses of techniques inspired from evolutionary biology such as selection, mutation, inheritance and recombination to solve a problem. The most commonly employed method in genetic algorithms is to create a group of individuals randomly from a given population. The individuals thus formed are evaluated with the help of the evaluation function provided by the programmer. Individuals are then provided with a score which indirectly highlights the fitness to the given situation. The best two individuals are then used to create one or more offspring, after which random mutations are done on the offspring. Depending on the needs of the application, the procedure continues until an acceptable solution is derived or until a certain number of generations have passed.

A genetic algorithm differs from a classical, derivative-based, optimization algorithm in two ways.

- A genetic algorithm generates a population of points in each iteration, whereas a classical algorithm generates a single point in each iteration.
- A genetic algorithm selects the next population by computation using random number generators, whereas a classical algorithm selects the next point by deterministic computation.

Compared to traditional artificial intelligence, a genetic algorithm provides many advantages. It is more robust and is not susceptible to breakdowns due to slight changes in inputs or due to the presence of noise.

Genetic algorithms are widely used in many fields such as robotics, automotive design, optimized telecommunications routing, engineering design and computer-aided design.

Unit 5

Text A

Expert System

扫码听课文

1. What is an expert system

An expert system is a computer program that is designed to solve complex problems and to provide decision-making ability like a human expert. It performs this by extracting knowledge from its knowledge base using the reasoning and inference rules according to the user queries.

The expert system is a part of AI, and the first ES was developed in the year 1970, which was the first successful approach of artificial intelligence. It solves the most complex issue as an expert by extracting the knowledge stored in its knowledge base. The system helps in decision-making using both facts and heuristics like a human expert. It is called so because it contains the expert knowledge of a specific domain and can solve any complex problem of that particular domain. The system is designed for a specific domain, such as medicine, science, etc.

The performance of an expert system is based on the expert's knowledge stored in its knowledge base. The more knowledge stored in the KB, the more that system improves its performance. One of the common examples of an ES is a suggestion of spelling errors while typing in the Google search box.

2. Characteristics of expert system

High performance: The expert system provides high performance for solving any type of complex problem of a specific domain with high efficiency and accuracy.

Understandable: It responds in a way that can be easily understood by the user. It can take input in human language and provides the output in the same way.

Reliable: It is much reliable for generating an efficient and accurate output.

Highly responsive: ES provides the result for any complex query within a very short period of time.

3. Components of expert system

An expert system mainly consists of three components: user interface, inference engine and knowledge base.

(1) User interface

With the help of a user interface, the expert system interacts with the user, takes queries as an input in a readable format, and passes it to the inference engine. After getting the response from the inference engine, it displays the output to the user. In other words, it is an interface that helps a non-expert user to communicate with the expert system to find a solution.

(2) Inference engine (rules of engine)

The inference engine is known as the brain of the expert system as it is the main processing unit of the system. It applies inference rules to the knowledge base to derive a conclusion or deduce new information. It helps in deriving an error-free solution of queries asked by the user. With the help of an inference engine, the system extracts the knowledge from the knowledge base.

There are two types of inference engine.

Deterministic inference engine: The conclusions drawn from this type of inference engine are assumed to be true. It is based on facts and rules.

Probabilistic inference engine: This type of inference engine contains uncertainty in conclusions, and it is based on the probability.

Inference engine uses the following two modes to derive the solutions.

Forward chaining: It starts from the known facts and rules, and applies the inference rules to add their conclusion to the known facts.

Backward chaining: It is a backward reasoning method that starts from the goal and works backward to prove the known facts.

(3) Knowledge base

The knowledge base is a type of storage that stores knowledge acquired from different experts of a particular domain. It is considered as big storage of knowledge. The bigger the knowledge base, the more precise the expert system will be. It is similar to a database that contains information and rules of a particular domain or subject. One can also view the knowledge base as collections of objects and their attributes. Such as a lion is an object and its attributes are it is a mammal, it is not a domestic animal, etc.

Components of knowledge base are as follows.

Factual knowledge: The knowledge which is based on facts and accepted by knowledge engineers comes under factual knowledge.

Heuristic knowledge: This knowledge is based on practice, the ability to guess, evaluation and experiences.

Knowledge representation: It is used to formalize the knowledge stored in the knowledge base using the if-else rules.

Knowledge acquisitions: It is the process of extracting, organizing and structuring the domain knowledge, specifying the rules to acquire the knowledge from various experts, and store that knowledge into the knowledge base.

4. Development of expert system

Here, we will explain the working of an expert system by taking an example of MYCIN ES. Below are some steps to build an MYCIN.

- Firstly, ES should be fed with expert knowledge. In the case of MYCIN, human experts specialized in the medical field of bacterial infection provide information about the causes, symptoms and other knowledge in that domain.

- The KB of the MYCIN is updated successfully. In order to test it, the doctor provides a new problem to it. The problem is to identify the presence of the bacteria by inputting the details of a patient, including the symptoms, current condition and medical history.

- The ES will need a questionnaire to be filled by the patient to know the general information about the patient, such as gender, age, etc.

- Now the system has collected all the information, so it will find the solution for the problem by applying if-then rules using the inference engine and using the facts stored within the KB.

- In the end, it will provide a response to the patient by using the user interface.

There are three primary participants in the building of expert system.

- Expert: The success of an ES depends much on the knowledge provided by human experts. These experts are those persons who are specialized in that specific domain.

- Knowledge engineer: Knowledge engineer is the person who gathers the knowledge from the domain experts and then codifies that knowledge to the system according to the formalism.

- End user: This is a particular person or a group of people who may not be experts, but work on the expert system to find the solution or advice for his queries.

5. Why expert system

Before using any technology, we must have an idea about why to use that technology and hence the same for the ES. The following describes the need of ES.

- No memory limitations: It can store as much data as required and can memorize it at the time of its application. But for human experts, there are some limitations to memorize all things at every time.

- High efficiency: If the knowledge base is updated with the correct knowledge, then it provides a highly efficient output, which may not be possible for a human.

- Expertise in a domain: There are lots of human experts in each domain, and they all have different experiences and different skills, so it is not easy to get a final output for the query. But if we put the knowledge gained from human experts into the expert system, It will provide an efficient output by mixing all the facts and knowledge.

- Not affected by emotions: These systems are not affected by human emotions such as fatigue, anger, depression, anxiety, etc. Hence the performance remains constant.

- High security: These systems provide high security to resolve any query.

- Considering all the facts: To respond to any query, it checks and considers all the available facts and provides the result accordingly. But it is possible that a human expert may not consider some facts due to any reason.

- Regular updates improve the performance: If there is an issue in the result provided by the expert systems, we can improve the performance of the system by updating the knowledge base.

6. Capabilities of expert system

- Advising: It is capable of advising the human being for the query of any domain from the particular ES.

- Providing decision-making capabilities: It provides the capability of decision making in any domain, such as for making any financial decision, decisions in medical science, etc.

- Demonstrating a device: It is capable of demonstrating any new products such as its features, specifications, how to use that product, etc.
- Problem-solving: It has problem-solving capabilities.
- Explaining a problem: It is also capable of providing a detailed description of an input problem.
- Interpreting the input: It is capable of interpreting the input given by the user.
- Predicting results: It can be used for the prediction of a result.
- Diagnosis: An ES designed for the medical field is capable of diagnosing a disease without using multiple components as it already contains various inbuilt medical tools.

7. Advantages of expert system

- These systems are highly reproducible.
- They can be used for risky places where the human presence is not safe.
- Error possibilities are less if the KB contains correct knowledge.
- The performance of these systems remains steady as it is not affected by emotions, tension or fatigue.
- They provide a very high speed to respond to a particular query.

8. Limitations of expert system

- The response of the expert system may get wrong if the knowledge base contains the wrong information.
- It cannot produce a creative output for different scenarios.
- Its maintenance and development costs are very high.
- Knowledge acquisition for designing is very difficult.
- For each domain, we require a specific ES, which is one of the big limitations.
- It cannot learn from itself and hence requires manual updates.

9. Applications of expert system

- In designing and manufacturing domain, it can be broadly used for designing and manufacturing physical devices such as camera lenses and automobiles.
- In the knowledge domain, these systems are primarily used for publishing the relevant knowledge to the users. The two popular expert systems used for this domain are an advisor and a tax advisor.
- In the finance industries, it is used to detect any type of possible fraud, suspicious activity, and advise bankers that if they should provide loans for business or not.
- In medical diagnosis, ES is used, and it was the first area where these systems were used.
- Expert systems can also be used for planning and scheduling some particular tasks for achieving the goal of that task.

✍ New Words

expert	['eksp3:t]	n.专家，行家
		adj.行家的；专业的

program	['prəʊɡræm]	n.程序
		v.给……编写程序
complex	['kɒmpleks]	adj.复杂的
decision-making	[dɪ'sɪʒn meɪkɪŋ]	n.决策
extract	['ekstrækt]	v.提取，提炼
inference	['ɪnfərəns]	n.推理；推断；推论
medicine	['medsn]	n.医学；药物
suggestion	[sə'dʒestʃən]	n.建议；表明
characteristic	[ˌkærəktə'rɪstɪk]	n.特色；特点
efficiency	[ɪ'fɪʃnsi]	n.效率，效能
accuracy	['ækjərəsi]	n.精确（性），准确（性）
understandable	[ˌʌndə'stændəbl]	adj.能懂的，可理解的
respond	[rɪ'spɒnd]	v.响应；回答，回复
input	['ɪnpʊt]	n.&v.输入
output	['aʊtpʊt]	n.&v.输出
reliable	[rɪ'laɪəbl]	adj.可信赖的；可靠的
responsive	[rɪ'spɒnsɪv]	adj.响应的，应答的
format	['fɔ:mæt]	n.格式
		vt.使格式化
display	[dɪ'spleɪ]	v.显示
derive	[dɪ'raɪv]	v.（使）起源于，来自；获得
deterministic	[dɪˌtɜːmɪ'nɪstɪk]	adj.确定性的
assume	[ə'sju:m]	v.假设，假定
uncertainty	[ʌn'sɜːtnti]	n.不确定；不可靠
collection	[kə'lekʃn]	n.集；群
practice	['præktɪs]	n.实践
development	[dɪ'veləpmənt]	n.开发；研制
infection	[ɪn'fekʃn]	n.传染，感染
symptom	['sɪmptəm]	n.症状；征兆
identify	[aɪ'dentɪfaɪ]	vt.识别，认出；确定
questionnaire	[ˌkwestʃə'neə]	n.调查表；调查问卷
codify	['kəʊdɪfaɪ]	v.编纂；整理；将（法规等）整理成典
describe	[dɪ'skraɪb]	v.描述；把……称为
emotion	[ɪ'məʊʃn]	n.情绪
anxiety	[æŋ'zaɪəti]	n.忧虑，焦虑
resolve	[rɪ'zɒlv]	vi.解决

regular	['reɡjələ]	*adj.*有规律的；定期的
advise	[əd'vaɪz]	*v.*劝告；建议
specification	[ˌspesɪfɪ'keɪʃn]	*n.*规格；详述；说明书
diagnosis	[ˌdaɪəɡ'nəʊsɪs]	*n.*诊断；判断
disease	[dɪ'ziːz]	*n.*疾病
inbuilt	['ɪnbɪlt]	*adj.*嵌入的，内置的
reproducible	[ˌriːprə'djuːsəbl]	*adj.*可再生的，可复写的；能繁殖的
tension	['tenʃn]	*n.*焦虑；冲突
fatigue	[fə'tiːɡ]	*n.*疲劳，厌倦
creative	[kri'eɪtɪv]	*adj.*创造性的，有创意的，创新的
scenario	[sə'nɑːriəʊ]	*n.*情景；设想
manual	['mænjuəl]	*adj.*用手的；手动的，手工的
		*n.*手册；指南
relevant	['reləvənt]	*adj.*有关的；相关联的
advisor	[əd'vaɪzə]	*n.*顾问
suspicious	[sə'spɪʃəs]	*adj.*可疑的；不信任的

✍ Phrases

store in	存储于，存储在
be designed for	为……而设计
search box	搜索框，搜索栏
period of time	时段，一段时间
consist of	包含；由……组成
user interface	用户界面
interact with	与……相互作用，与……相互影响，与……相互配合
main processing unit	主处理单元，主处理机
deterministic inference engine	确定性推理引擎
probabilistic inference engine	概率推理引擎
forward chaining	前向链接
backward chaining	后向链接
factual knowledge	事实知识
knowledge acquisition	知识获得，知识收集
end user	最终用户
a group of	一群，一组

✎ Abbreviations

ES (Expert System)	专家系统
KB (Knowledge Base)	知识库

Reference Translation

专家系统

1. 什么是专家系统

专家系统是一种计算机程序，旨在像人类专家一样解决复杂问题并提供决策能力。它通过根据用户查询使用推理和推理规则，从知识库中提取知识来执行此操作。

专家系统是人工智能的一部分，第一个专家系统于 1970 年开发，这是第一个成功的人工智能方法。它像专家一样通过提取存储在知识库中的知识来解决最复杂的问题。该系统像人类专家一样使用事实和启发式方法帮助做出决策。之所以这样称呼，是因为它包含特定领域的专家知识，并且可以解决该特定领域的任何复杂问题。该系统是为特定领域设计的，如医学、科学等。

专家系统的性能基于存储在其知识库中的专家的知识。知识库中存储的知识越多，系统性能就越好。专家系统的常见例子之一是在 Google 搜索框中键入时提示拼写错误。

2. 专家系统的特点

高性能：专家系统可高效、高精度地解决特定领域的任何类型的复杂问题。

可以理解：它以用户易于理解的方式进行响应。它可以接受人类语言的输入，并以相同的方式提供输出。

可靠：非常可靠地生成高效、准确的输出。

响应速度快：专家系统可以在很短的时间内为任何复杂的查询提供结果。

3. 专家系统的组成

专家系统主要由用户界面、推理引擎和知识库三部分组成。

（1）用户界面

在用户界面的帮助下，专家系统与用户进行交互，以可读格式将查询作为输入，并将其传递给推理引擎。从推理引擎获得响应后，它会将输出显示给用户。换句话说，它是一个接口，帮助非专家用户与专家系统进行通信以找到解决方案。

（2）推理引擎（引擎规则）

推理引擎被称为专家系统的大脑，因为它是系统的主要处理单元。它将推理规则应用于知识库，以得出结论或推论新信息。它有助于得出用户询问的无错误解决方案。在推理引擎的帮助下，系统从知识库中提取知识。

有两种类型的推理引擎。

- 确定性推理引擎：假定从这种类型的推理引擎得出的结论是正确的。它基于事实和规则。

- 概率推理引擎：这种类型的推理引擎结论中包含不确定性，它是基于概率的。

推理引擎使用以下两种模式来得出解决方案。

正向链接：它从已知事实和规则开始，并应用推理规则将其结论添加到已知事实中。

反向链接：这是一种从目标开始并反向证明已知事实的反向推理方法。

（3）知识库

知识库是一种存储类型，用于存储从特定领域的不同专家那里获得的知识。它被认为是大量的知识存储。知识库越大，专家系统越精确。它类似于包含特定领域或主题的信息和规则的数据库。人们还可以将知识库视为对象及其属性的集合。例如狮子是物体、它的属性是哺乳动物、不是家畜等。

知识库的组成部分如下。

事实知识：基于事实并被知识工程师接受的知识属于事实知识。

启发式知识：该知识基于实践、猜测能力、评估和经验。

知识表示：它用于使用 if-else 规则来形式化存储在知识库中的知识。

知识获取：这是提取、组织和构造领域知识并指定规则，以从各种专家那里获取知识并将该知识存储到知识库中的过程。

4. 专家系统的开发

在这里，我们将以 MYCIN 专家系统为例来说明专家系统的工作原理。以下是构建 MYCIN 的一些步骤。

- 首先，应向专家系统提供专业知识。就 MYCIN 而言，专门从事细菌感染医学领域的人类专家将提供有关该领域的原因、症状和其他知识的信息。
- MYCIN 的知识库已成功更新。为了测试它，医生给它提供了一个新的问题。该问题是通过输入患者的详细信息（包括症状、当前状况和病史）来辨别细菌的存在。
- 专家系统让需要的患者填写问卷，以了解有关患者的一般信息，如性别、年龄等。
- 现在系统已经收集了所有的信息，通过推理引擎应用 if-then 规则，利用知识库中存储的事实，就可以找到问题的解决方案。
- 最后，它将通过用户界面为患者提供响应。

专家系统的构建有三个主要参与者。

- 专家：专家系统的成功很大程度上取决于人类专家提供的知识。这些专家是专门从事特定领域的那些人。
- 知识工程师：知识工程师是从领域专家那里收集知识，然后根据形式体系将知识编码到系统中的人。
- 最终用户：这是特定的一个人或一群人，他们可能不是专家，但在专家系统上工作，从而为他的查询找到解决方案或建议。

5. 为什么选择专家系统

在使用任何技术之前，我们必须先了解为什么要使用该技术，因此对于专家系统也是一样。以下描述了专家系统的需求。

- 没有内存限制：它可以按需存储数据，并可以在应用时记住这些数据。但是对于人类专家而言，每次记住所有东西都有一定的局限性。
- 高效：如果用正确的知识更新知识库，那么它将提供高效的结果，这对于人类而言也许是不可能的。

- 某个领域的专业知识：每个领域都有很多人类专家，他们都有不同的经验和不同的技能，因此要获得查询的最终结果并不容易。但是，如果我们把从人类专家那里获得的知识放到专家系统中，它将通过混合所有事实和知识来提供有效的结果。
- 不受情感影响：这些系统不受诸如疲劳、愤怒、沮丧、焦虑等人类情感的影响。因此，性能可以保持不变。
- 高安全性：这些系统为解决任何查询提供了高度的安全性。
- 考虑所有事实：为了响应任何查询，它会检查并考虑所有现有事实，并相应地提供结果。但是，人类专家有可能由于任何原因而没有考虑某些事实。
- 定期更新可以提高性能：如果专家系统提供的结果存在问题，我们可以通过更新知识库来提高系统的性能。

6. 专家系统的能力

- 建议：它可以建议人类从特定专家系统查询任何领域。
- 提供决策能力：它提供任何领域的决策能力，如做出任何财务决策、医学决策等。
- 演示设备：它能够演示任何新产品，如其功能、规格、如何使用该产品等。
- 解决问题：它具有解决问题的能力。
- 解释问题：它也能够提供输入问题的详细描述。
- 解释输入：能够解释用户给出的输入。
- 预测结果：它可用于预测结果。
- 诊断：专为医疗领域设计的专家系统无须使用多个组件即可诊断疾病，因为它已经包含了各种内置医疗工具。

7. 专家系统的优点

- 这些系统具有很高的可重复性。
- 它们可用于不保证人身安全的危险场所。
- 如果知识库包含正确的知识，则出错的可能性会较小。
- 由于不受情绪、紧张或疲劳的影响，这些系统的性能保持稳定。
- 对特定查询，它们提供了很高的响应速度。

8. 专家系统的局限性

- 如果知识库包含错误的信息，则专家系统的响应可能会出错。
- 它无法针对不同场景产生创意输出。
- 其维护和开发成本很高。
- 设计方面的知识获取非常困难。
- 对于每个领域，我们都需要特定的专家系统，这是最大的限制之一。
- 它无法从自身学习，因此需要手动更新。

9. 专家系统的应用

- 在设计和制造领域，它可以广泛地用于设计和制造相机镜头与汽车等物理设备。
- 在知识领域，这些系统主要用于将相关知识发布给用户。用于此领域的两个流行的专家系统是顾问和税务顾问。
- 在金融行业中，它用于检测任何类型的可能的欺诈、可疑活动，并建议银行家是否应该为企业提供贷款。

- 在医疗诊断中使用了专家系统，这是使用这些系统的第一个领域。
- 专家系统还可以用于计划和安排某些特定任务，以实现该任务的目标。

Text B

扫码听课文

Computer Vision

1. What is Computer Vision (CV)

Computer vision is a field of study which enables computers to replicate the human visual system. It's a subset of artificial intelligence which collects information from digital images or videos and processes them. The entire process involves image acquiring, screening, analysing, identifying and extracting information. This extensive processing helps computers to understand any visual content and act on it accordingly.

Computer vision projects translate digital visual content into explicit descriptions to gather multi-dimensional data. This data is then turned into computer-readable language to aid the decision-making process. The main objective of this branch of artificial intelligence is to teach machines to collect information from pixels.

2. Examples of computer vision and algorithms

Automatic cars aim at reducing the need for human intervention while driving, through various AI systems. Computer vision is part of such a system which focuses on imitating the logics behind human vision to help the machines take data-based decisions. CV systems will scan live objects and categorise them, based on which the car will keep running or make a stop. If the car comes across an obstacle or a traffic light, it will analyse the image, create a 3D version of it, consider the features and decide on an action—all within a second.

3. How does computer vision work

Computer vision primarily relies on pattern recognition techniques to self-train and understand visual data. The wide availability of data and the willingness of companies to share them have made it possible for deep learning experts to use this data to make the process more accurate and fast.

While machine learning algorithms were previously used for computer vision applications, now deep learning methods have evolved as a better solution for this domain. For instance, machine learning techniques require a humongous amount of data and active human monitoring in the initial phase to ensure that the results are as accurate as possible. Deep learning, on the other hand, relies on neural networks and uses examples for problem solving. It self-learns by using labeled data to recognise common patterns in the examples.

4. Why is computer vision important

From selfies to landscape images, we are flooded with all kinds of photos today. According to a report by Internet Trends, people upload more than 1.8 billion images every day, and that's just the number of uploaded images. Imagine what the number would come to if you consider the images stored in phones. We consume more than 4,146,600 videos on YouTube and send 103,447,520 spam mails everyday. Again, that's just a part of it—communication, media and entertainment, the internet of things are all actively contributing to this number. This abundantly available visual content demands analysing and understanding. Computer vision helps in doing that by teaching machines to "see" these images and videos.

Additionally, thanks to easy connectivity, the internet is easily accessible by all today. Children are especially susceptible to online abuse and "toxicity". Apart from automating a lot of functions, computer vision also ensures moderation and monitoring of online visual content. One of the main tasks involved in online content curation is indexing. The content available on the internet is mainly of several types, namely text, image, video and audio. Computer vision uses algorithms to read and index images. Popular search engines like Google and Youtube use computer vision to scan through images and videos to approve them for featuring. By way of doing so, they not only provide users with relevant content but also protect against online abuse and "toxicity".

5. Which language is best suited for computer vision

We have several programming languages to choose for computer vision—OpenCV using C++, OpenCV using Python or MATLAB. However, most engineers have a personal favourite, depending on the task they perform. Beginners often pick OpenCV with Python for its flexibility. It's a language most programmers are familiar with, and owing to its versatility it is very popular among developers.

Computer vision experts recommend Python for the following reasons.

- Easy to use: Python is easy to learn, especially for beginners. It is one of the first programming languages learnt by most users. This language is also easily adaptable for all kinds of programming needs.
- Most used computing language: Python offers a complete learning environment for people who want to use it for various kinds of computer vision and machine learning experiments. Its NumPy, Scikit-learn, Matplotlib and OpenCV provide an exhaustive resource for any computer vision applications.
- Debugging and visualisation: Python has an inbuilt debugger, "PDB", which makes debugging codes in this programming language more accessible. Similarly, Matplotlib is a convenient resource for visualisation.
- Web backend development: Frameworks like Django, Flask, and Web2py are excellent web page builders. Python is compatible with these frameworks and can be easily tweaked to fit your requirements.

MATLAB is a software package popular with computer experts. Let's look into the advantages of using MATLAB.

- Toolboxes: MATLAB has very exhaustive toolboxes. Whether it is a statistical and machine learning toolbox, or an image processing toolbox, MATLAB has one included for all kinds of needs. The clean interfaces of each of these toolboxes enable you to implement a range of algorithms. MATLAB also has an optimisation toolbox which ensures that all algorithms perform at their best.

- Powerful matrix library: Images and other visual content contains multi-dimensional matrices along with linear algebra in different algorithms, which makes it easier to work within MATLAB. The linear algebra routines included in MATLAB work fast and effective.

- Debugging and visualisation: Since there is a single integrated platform for coding in MATLAB, writing, visualising and debugging codes become easy.

- Excellent documentation: MATLAB enables you to document your work adequately so that it is accessible later. Documentation is essential not just for future reference but also to help coders work faster. MATLAB's documentation allows users to work twice the speed of OpenCV.

6. Applications of computer vision

- Medical imaging: Computer vision helps in MRI reconstruction, automatic pathology, diagnosis, machine aided surgeries and more.

- AR/VR: Object occlusion (dense depth estimation), outside-in tracking, inside-out tracking for virtual and augmented reality.

- Smartphones: All the photo filters (including animation filters on social media), QR code scanners, panorama construction, computational photography, face detectors, image detectors (Google Lens, Night Sight) that you use are computer vision applications.

- Internet: Image search, geolocalisation, image captioning, imaging for maps, video categorisation and more.

7. Computer vision challenges

Computer vision might have emerged as one of the top fields of machine learning, but there are still several obstacles in its way of becoming a leading technology. Human vision is a complicated and highly effective system which is difficult to replicate through technology. However, that's not to say that computer vision will not improve in the future.

The challenges we face in computer vision are as follows.

- Reasoning issue: Modern neural network-based algorithms are complex systems whose functionings are often obscure. In situations like these, it becomes tough to find the logic behind any task. This lack of reasoning creates a real challenge for computer vision experts who try to define any attribute in an image or video.

- Privacy and ethics: Vision powered surveillance is a serious threat to privacy in a lot of countries. It exposes people to unauthorised use of data. Face recognition and detection is prohibited in some countries because of these problems.

- Fake content: Like all other technologies, computer vision in the wrong hands can lead to dangerous problems. Anybody with access to powerful data centers is capable of creating fake images, videos or text content.

8. Future of computer vision

Computer vision is a fast-developing field and has gathered a lot of attention from various industries. It will be able to function on a broader spectrum of content in the future. With the amount of data we are generating every day, it's only natural that machines will use that data to craft solutions.

Once computer vision experts can resolve the current problems of the domain, we can expect a trustworthy system that automates content moderation and monitoring.

✎ New Words

vision	['vɪʒn]	n.视觉，视力
enable	[ɪ'neɪbl]	v.使能够；使可行
subset	['sʌbset]	n.子集
screen	[skriːn]	v.筛选
visual	['vɪʒuəl]	adj.视觉的
translate	[trænz'leɪt]	v.（使）转变为
explicit	[ɪk'splɪsɪt]	adj.易于理解的；明确的
multi-dimensional	[,mʌltɪ daɪ'menʃənl]	adj.多维的；多重的
pixel	['pɪksl]	n.像素
data-based	['deɪtə beɪst]	adj.以数据为基础，基于数据（的）
scan	[skæn]	v.扫描
willingness	['wɪlɪŋnəs]	n.愿意，乐意
share	[ʃeə]	v.共有，合用，分享
humongous	[hjuː'mʌŋgəs]	adj.极大的，奇大无比的
active	['æktɪv]	adj.积极的；活跃的
upload	[,ʌp'ləʊd]	vt.上传，上载
media	['miːdɪə]	n.媒体
abundantly	[ə'bʌndəntli]	adv.丰富地；大量地
video	['vɪdiəʊ]	n.视频
susceptible	[sə'septəbl]	adj.易受影响的
abuse	[ə'bjuːs]	n.&v.滥用
moderation	[,mɒdə'reɪʃn]	n.适度；自我节制；稳定
curation	[,kjuː'reɪʃn]	n.管理；综合处理
protect	[prə'tekt]	v.保护
flexibility	[,fleksə'bɪləti]	n.灵活性；弹性
versatility	[,vɜːsə'tɪləti]	n.多用途性
developer	[dɪ'veləpə]	n.（产品等的）开发者
exhaustive	[ɪg'zɔːstɪv]	adj.详尽的，彻底的
resource	[rɪ'sɔːs]	n.资源
debug	[,diː'bʌg]	vt.调试；排除故障

debugger	[ˌdiːˈbʌgə]	n.调试器，调试程序
convenient	[kənˈviːniənt]	adj.实用的；方便的
framework	[ˈfreɪmwɜːk]	n.构架；框架；（体系的）结构
builder	[ˈbɪldə]	n.构建器；开发器
compatible	[kəmˈpætəbl]	adj.兼容的
optimisation	[ˌɒptɪmaɪˈzeɪʃən]	n.最优法；最优化
ensure	[ɪnˈʃʊə]	v.确保；担保
matrix	[ˈmeɪtrɪks]	n.矩阵
routine	[ruːˈtiːn]	n.例程；惯例，常规 adj.常规的，日常的，平常的
integrate	[ˈɪntɪgreɪt]	v.合并；集成
platform	[ˈplætfɔːm]	n.平台
code	[kəʊd]	n.代码；编码 v.把……编码；编程序
reference	[ˈrefrəns]	v.引用 n.提及；查询；征求 adj.供参考的
coder	[ˈkəʊdə]	n.程序员，编程者；编码器
reconstruction	[ˌriːkənˈstrʌkʃn]	n.重建，再现；重建物，复原物
estimation	[ˌestɪˈmeɪʃn]	n.评价，判断；估算
filter	[ˈfɪltə]	n.过滤器；筛选（过滤）程序 v.过滤
scanner	[ˈskænə]	n.扫描器；扫描设备
geolocalisation	[ˈdʒiːəʊˌləʊkəlaɪˈzeɪʃn]	n.地理定位
categorisation	[ˌkætɪgəraɪˈzeɪʃn]	n.分类
complicate	[ˈkɒmplɪkeɪt]	v.使复杂化
obscure	[əbˈskjʊə]	adj.不清楚的；隐蔽的 vt.使难理解；掩盖；隐藏
ethic	[ˈeθɪk]	n.道德规范；伦理标准
surveillance	[sɜːˈveɪləns]	n.盯梢，监督；监视
serious	[ˈsɪəriəs]	adj.严重的；令人担忧的
threat	[θret]	n.威胁
expose	[ɪkˈspəʊz]	n.暴露，揭露
prohibit	[prəˈhɪbɪt]	v.禁止，阻止
dangerous	[ˈdeɪndʒərəs]	adj.危险的
spectrum	[ˈspektrəm]	n.光谱，波谱；范围；系列
craft	[krɑːft]	v.精心制作 n.手艺，技巧

| trustworthy | ['trʌstwɜːði] | *adj.*值得信赖的，可靠的 |

Phrases

digital image	数字图像
act on	按照……而行动；遵行
multi-dimensional data	多维数据
computer-readable language	计算机可读语言
focus on	聚焦
traffic light	红绿灯，交通灯
rely on	依靠，依赖
pattern recognition	模式识别
visual data	可视化数据
machine learning algorithm	机器学习算法
initial phase	初起阶段
labeled data	已标记的数据
flooded with	淹没；挤满，充满
spam mail	垃圾邮件
internet of things	物联网
contribute to	贡献；有助于
personal favourite	个人喜好
be adaptable for	适合于
web page	网页
be compatible with	与……兼容
a range of	一系列；一些；一套
linear algebra	线性代数
machine aided surgery	机器辅助手术
object occlusion	目标遮挡
dense depth estimation	密集深度估计
social media	社交媒体
QR code	二维码
panorama construction	全景建筑
face detector	人脸检测器

Abbreviations

| CV (Computer Vision) | 计算机视觉 |
| MRI (Magnetic Resonance Imaging) | 磁共振成像 |

AR (Augmented Reality) 　　　　　增强现实
VR (Virtual Reality) 　　　　　　虚拟现实

Reference Translation

计算机视觉

1. 什么是计算机视觉（CV）

计算机视觉是使计算机能够复制人类视觉系统的研究领域。它是人工智能的子集，可以从数字图像或视频中收集信息并对其进行处理。整个过程涉及图像获取、筛选、分析、识别和提取信息。这种广泛的处理过程可以帮助计算机理解任何视觉内容并据此采取行动。

计算机视觉项目将数字视觉内容转换为明确的描述，以收集多维数据。然后将这些数据转换为计算机可读的语言，以辅助决策过程。人工智能这一分支的主要目的是教机器从像素中收集信息。

2. 计算机视觉和算法示例

自动驾驶汽车旨在通过各种人工智能系统减少驾车时人为干预的需求。计算机视觉是这种系统的一部分，该系统专注于模仿人类视觉背后的逻辑，以帮助机器做出基于数据的决策。计算机视觉系统将扫描活动物体并对其进行分类，汽车将以此为基础继续行驶或停车。如果汽车遇到障碍物或交通信号灯，它将分析图像，创建图像的三维版本，考虑其功能并决定要采取的措施——所有这些操作都将在一秒钟之内完成。

3. 计算机视觉如何工作

计算机视觉主要依靠模式识别技术来自我训练和理解视觉数据。数据的广泛可用性以及公司共享数据的意愿使深度学习专家可以使用此数据来使过程更加准确和快速。

虽然机器学习算法以前用于计算机视觉应用，但如今，深度学习方法已经发展成为该领域的一种更好的解决方案。例如，机器学习技术在初始阶段需要大量数据和积极的人工监管，以确保结果尽可能准确。而深度学习依赖于神经网络并使用示例来解决问题。它通过使用标记的数据来识别示例中的常见模式来进行自学习。

4. 为什么计算机视觉很重要

从自拍照到风景图像，如今到处都是各种各样的照片。根据 Internet Trends 的报告，人们每天上传超过 18 亿张图片，而这仅仅是上传图片的数量。想象一下，如果考虑手机中存储的图像，数量将是多少。我们每天在 YouTube 上观看超过 4 146 600 个视频，发送103 447 520 封垃圾邮件。同样，这只是其中的一部分——通信、媒体和娱乐、物联网都在为这一数字做出积极贡献。这种大量可用的视觉内容需要分析和理解。计算机视觉可以通过教机器"查看"这些图像和视频来帮助实现这一目标。

此外，得益于便捷的连接性，如今所有人都可以轻松访问互联网。儿童特别容易受到在线滥用和"毒性"的影响。除了自动执行许多功能外，计算机视觉还可以确保对在线视觉内容进行审核和监控。在线内容管理所涉及的主要任务之一是索引编制。互联网上现有的内容主要有几种类型，即文本、图像、视频和音频。计算机视觉使用算法来读取和索引图像。诸如 Google 和 Youtube 之类的流行搜索引擎使用计算机视觉来扫描图像和视频以批准其功能。这样，它们不仅为用户提供相关内容，而且还可以防止在线滥用和"毒性"。

5. 哪种语言最适合计算机视觉

对于计算机视觉，我们有多种编程语言可供选择——使用 C ++的 OpenCV，使用 Python 的 OpenCV 或 MATLAB。但是，大多数工程师会根据自己执行的任务而对其进行个性化设置。初学者经常选择使用 Python 的 OpenCV 来提高灵活性，它是大多数程序员都熟悉的一种语言，并且由于其多功能性，在开发人员中非常受欢迎。

计算机视觉专家推荐 Python 的原因如下。

- 易于使用：Python 易于学习，特别是对于初学者。它是大多数用户最早学习的编程语言之一。这种语言也很容易适应各种编程需求。
- 最常用的计算语言：Python 为想要将其用于各种计算机视觉和机器学习实验的人们提供了一个完整的学习环境。它的 NumPy、Scikit-learn、Matplotlib 和 OpenCV 为所有计算机视觉应用程序提供了详尽的资源。
- 调试和可视化：Python 具有内置的调试器"PDB"，可以更轻松地访问此编程语言中的调试代码。同样，Matplotlib 是可视化的便捷资源。
- Web 后端开发：Django、Flask 和 Web2py 之类的框架是出色的网页构建器。Python 与这些框架兼容，并且可以轻松地进行调整以满足你的要求。

MATLAB 是一个受计算机专家欢迎的软件包。让我们来看看使用 MATLAB 的优势。

- 工具箱：MATLAB 具有非常详尽的工具箱。无论是统计和机器学习工具箱，还是图像处理工具箱，MATLAB 都有满足所有需求的工具箱。每个工具箱的简洁界面使你可以实现一系列算法。MATLAB 还有一个优化工具箱，可确保所有算法都能发挥最佳性能。
- 强大的矩阵库：图像和其他视觉内容包含多维矩阵以及不同算法中的线性代数，这使得在 MATLAB 中更容易工作。MATLAB 中包含的线性代数例程可以快速而有效地工作。
- 调试和可视化：由于在 MATLAB 中有一个用于编码的集成平台，因此编写、可视化和调试代码变得容易。
- 出色的文档编制能力：MATLAB 使你能够充分地记录你的工作，以便以后可以访问。文档不仅对于将来的参考至关重要，而且对于帮助编码人员更快地工作也至关重要。MATLAB 的文档使用户可以以两倍于 OpenCV 的速度工作。

6. 计算机视觉的应用

- 医学成像：计算机视觉有助于 MRI 再现、自动病理学、诊断、机器辅助手术等。
- 增强现实/虚拟现实：在虚拟和增强现实中，进行目标遮挡（密集深度估计）、从外到内跟踪和从内到外跟踪。
- 智能手机：你使用的所有照片滤镜（包括社交媒体上的动画滤镜）、二维码扫描仪、全景建筑、计算摄影、人脸检测器、图像检测器（Google Lens 及 Night Sight）都是计算机视觉应用。
- 互联网：图像搜索、地理定位、图像标题、地图成像、视频分类等。

7. 计算机视觉挑战

计算机视觉可能已经成为机器学习的顶级领域之一，但是在成为领先技术方面仍然存在一些障碍。人类的视觉是一个复杂而高效的系统，很难通过技术来复制。但是，这并不是说将来计算机视觉不会得到改善。

我们在计算机视觉中面临的挑战如下。

- 推理问题：基于现代神经网络的算法是复杂的系统，其功能通常难以理解。在这种情况下，很难找到任何任务背后的逻辑。这种缺乏推理的现象给试图定义图像或视频中任何属性的计算机视觉专家带来了真正的挑战。
- 隐私和道德：在许多国家，视觉监控对隐私构成严重威胁。它使人们暴露于未经授权的数据使用之下。由于这些问题，在某些国家/地区禁止脸部识别和检测。
- 虚假内容：与其他各种技术一样，不正确地使用计算机视觉可能会导致危险的问题。任何有权访问强大数据中心的人都可以创建伪造的图像、视频或文本内容。

8. 计算机视觉的未来

计算机视觉是一个快速发展的领域，并且受到了各个行业的广泛关注。将来它能够在更广泛的内容中发挥作用。随着我们每天生成的数据量的增加，机器自然会使用这些数据来制定解决方案。

一旦计算机视觉专家可以解决该领域的当前问题，我们就可以期待一个可信赖的系统，它可以自动进行内容审核和监控。

Exercises

[Ex. 1] Answer the following questions according to Text A.

1. What is an expert system? How does it perform this?

2. What is the performance of an expert system based on?

3. What are the characteristics of expert system?

4. What does an expert system mainly consist of?

5. Why is the inference engine known as the brain of the expert system? What are the two types of inference engine?

6. What are the components of knowledge base?

7. How many primary participants in the building of expert system are there? What are they?

8. What are the capabilities of expert system?

9. When may the response of the expert system get wrong ?

10. What is expert system used to do in the finance industries?

[Ex. 2] Fill in the following blanks according to Text B.

1. Computer vision is a field of study which enables ＿＿＿＿＿＿＿ to replicate ＿＿＿＿＿＿＿＿＿＿. It's a subset of ＿＿＿＿＿＿＿ which collects information from ＿＿＿＿＿＿ or videos and processes them.

2. Automatic cars aim at reducing the need for ＿＿＿＿＿＿＿ while driving, through ＿＿＿＿＿＿＿. Computer vision is part of such a system which focuses on ＿＿＿＿＿＿＿

behind human vision to help the machines _____ .

3. Computer vision primarily relies on _____ to self-train and understand _____. The _____ of data and the willingness of companies to _____ have made it possible for deep learning experts to use this data to make the process _____ .

4. From selfies to landscape images, we are flooded with _____ today. According to a report by Internet Trends, people upload more than _____ every day, and that's just the number of _____ .

5. Computer vision uses algorithms to _____ and _____ images. Popular search engines like _____ and _____ use computer vision to scan through images and videos to _____ .

6. We have several programming languages to choose for computer vision — _____ , _____ or _____ . However, most engineers have a personal favourite, depending on _____ .

7. Computer vision experts recommend Python for the following reasons: _____ , _____ , _____ and _____ .

8. The advantages of using MATLAB are _____ , _____ , _____ and _____ .

9. Computer vision is applied in _____ , _____ , _____ and _____ .

10. The challenges we face in computer vision are _____ , _____ and _____ .

[Ex. 3] Translate the following terms or phrases from English into Chinese.

1.	accuracy	1.	_____
2.	assume	2.	_____
3.	display	3.	_____
4.	extract	4.	_____
5.	inbuilt	5.	_____
6.	end user	6.	_____
7.	deterministic inference engine	7.	_____
8.	factual knowledge	8.	_____
9.	knowledge representation	9.	_____
10.	knowledge acquisition	10.	_____

[Ex. 4] Translate the following terms or phrases from Chinese into English.

1.	*n*.推理；推断；推论	1.	_____
2.	*n*.程序　*v*.给……编写程序	2.	_____
3.	*v*.响应；回答，回复	3.	_____

4. *n.*规格；详述；说明书   4. _____

5. *adj.*兼容的   5. _____

6. 概率推理引擎   6. _____

7. 用户界面   7. _____

8. 计算机可读语言   8. _____

9. 聚焦   9. _____

10. 物联网   10. _____

[Ex. 5] Translate the following passage into Chinese.

Decision Tree

1. What is a decision tree

A decision tree is a type of flowchart that shows a clear pathway to a decision. In terms of data analytics, it is a type of algorithm that includes conditional "control" statements to classify data. A decision tree starts at a single point (or "node") which then branches (or "splits") in two or more directions. Each branch offers different possible outcomes, incorporating a variety of decisions and chance events until a final outcome is achieved. When shown visually, their appearance is tree-like, hence the name.

Decision trees are extremely useful for data analytics and machine learning because they break down complex data into more manageable parts. They're often used for prediction analysis, data classification and regression.

2. Pros and cons of decision trees

Decision trees are very powerful tools. Nevertheless, like any algorithm, they're not suited to every situation. Here are some key advantages and disadvantages of decision trees.

The advantages of decision trees are as follows.

- Good for interpreting data in a highly visual way.
- Good for handling a combination of numerical and non-numerical data.
- Easy to define rules, e.g. "yes, no, if, then, else…".
- Requires minimal preparation or data cleaning before use.
- Great way to choose between best, worst and likely case scenarios.
- Can be easily combined with other decision-making techniques.

The disadvantages of decision trees are as follows.

- Overfitting (where a model interprets meaning from irrelevant data) can become a problem if a decision tree's design is too complex.
- They are not well-suited to continuous variables (i.e. variables which can have more than one value or a spectrum of values).
- In predictive analysis, calculations can quickly grow cumbersome, especially when a decision path includes many chance variables.
- When using an imbalanced dataset (i.e. where one class of data dominates over another) it is easy for outcomes to be biased in favor of the dominant class.

- Generally, decision trees provide lower prediction accuracy compared to other predictive algorithms.

3. What are decision trees used for

Despite their drawbacks, decision trees are still a powerful and popular tool. They're commonly used by data analysts to carry out predictive analysis (e.g. to develop operations strategies in businesses). They're also a popular tool for machine learning and artificial intelligence, where they're used as training algorithms for supervised learning.

Broadly, decision trees are used in a wide range of industries, to solve many types of problems. Because of their flexibility, they're used in sectors from technology and health to financial planning.

Unit 6

Text A

Machine Learning

扫码听课文

Machine learning is a branch of Artificial Intelligence (AI) focused on building applications that learn from data and improve their accuracy over time without being programmed to do so.

In data science, an algorithm is a sequence of statistical processing steps. In machine learning, algorithms are "trained" to find patterns and features in massive amounts of data in order to make decisions and predictions based on new data. The better the algorithm, the more accurate the decisions and predictions will become as it processes more data.

1. How machine learning works

There are four basic steps for building a machine learning application (or model). These are typically performed by data scientists working closely with business professionals for whom the model is being developed.

(1) Select and prepare a training data set

Training data is a data set representative of the data the machine learning model will ingest to solve the problem it's designed to solve. In some cases, the training data is labeled data—"tagged" to call out features and classifications the model will need to identify. Other data is unlabeled, and the model will need to extract those features and assign classifications on its own.

In either case, the training data needs to be properly prepared—randomized, deduped, and checked for imbalances or biases that could impact the training. It should also be divided into two subsets: the training subset, which is used to train the application, and the evaluation subset, which is used to test and refine it.

(2) Choose an algorithm to run on the training data set

Again, an algorithm is a set of statistical processing steps. The type of algorithm depends on the type (labeled or unlabeled) and amount of data in the training data set and on the type of problem to be solved.

Common types of machine learning algorithms with labeled data are as follows.

- Regression algorithms: Linear and logistic regression are examples of regression algorithms used to understand relationships in data. Linear regression is used to predict the value of a dependent variable based on the value of an independent variable. Logistic regression can be used when the dependent variable is binary in nature: A or B. For example, a linear regression algorithm could be trained to predict a salesperson's annual sales (the dependent variable) based on its relationship to the salesperson's education or years of experience (the independent variables). Another type of regression algorithm called a support vector machine is useful when dependent variables are more difficult to classify.

- Decision trees: Decision trees use classified data to make recommendations based on a set of decision rules. For example, a decision tree that recommends betting on a particular horse to win, place or show could use data about the horse (e.g., age, winning percentage, pedigree) and apply rules to those factors to recommend an action or decision.
- Instance-based algorithms: A good example of an instance-based algorithm is KNN. It uses classification to estimate how likely a data point is to be a member of one group or another based on its proximity to other data points.

Algorithms with unlabeled data include the following.

- Clustering algorithms: Think of clusters as groups. Clustering focuses on identifying groups of similar records and labeling the records according to the group to which they belong. This is done without prior knowledge about the groups and their characteristics. Types of clustering algorithms include the K-means, TwoStep and Kohonen clustering.
- Association algorithms: Association algorithms find patterns and relationships in data and identify frequent "if-then" relationships called association rules. These are similar to the rules used in data mining.
- Neural networks: A neural network is an algorithm that defines a layered network of calculations, It has an input layer (where data is ingested), at least one hidden layer (where calculations are performed and make different conclusions about input) and an output layer (where each conclusion is assigned a probability). A deep neural network defines a network with multiple hidden layers, each of which successively refines the results of the previous layer.

(3) Training the algorithm to create the model

Training the algorithm is an iterative process. It involves running variables through the algorithm, comparing the output with the results it should have produced, adjusting weights and biases within the algorithm that might yield a more accurate result, and running the variables again until the algorithm returns the correct result most of the time.

(4) Using and improving the model

The final step is to use the model with new data and to improve the model's accuracy and effectiveness over time. Where the new data comes from will depend on the problem being solved. For example, a machine learning model designed to identify spam will ingest email messages, whereas a machine learning model that drives a robot vacuum cleaner will ingest data resulting from real-world interaction with moved furniture or new objects in the room.

2. Machine learning methods

(1) Supervised machine learning

Supervised machine learning trains itself on a labeled data set. That is, the data is labeled with information that the machine learning model is being built and that may even be classified in ways the model is supposed to classify data. For example, a computer vision model designed to identify purebred German Shepherd dogs might be trained on a data set of various labeled dog images.

Supervised machine learning requires less training data than other machine learning methods and it makes training easier because the results of the model can be compared to actual labeled results. But, properly labeled data is expensive to prepare, and there's the danger of overfitting, or creating a model so closely tied and biased to the training data that it doesn't handle variations in new data accurately.

(2) Unsupervised machine learning

Unsupervised machine learning ingests lots and lots of unlabeled data. It uses algorithms to extract meaningful features of the data without human intervention. Unsupervised learning is less about automating decisions and predictions, and more about identifying patterns and relationships in data that humans would miss. Take spam detection for example, people generate more email than a team of data scientists could ever hope to label or classify in their lifetimes. An unsupervised learning algorithm can analyze huge volumes of emails and uncover the features and patterns that indicate spam (and keep getting better at flagging spam over time).

(3) Semi-supervised machine learning

Semi-supervised machine learning offers a good medium between supervised and unsupervised machine learning. During training, it uses a smaller labeled data set to guide classification and feature extraction from a larger, unlabeled data set. Semi-supervised learning can solve the problem of having not enough labeled data (or not being able to afford to label enough data) to train a supervised learning algorithm.

(4) Reinforcement machine learning

Reinforcement machine learning is a behavioral machine learning model that is similar to supervised learning, but the algorithm isn't trained using sample data. This model learns as it goes by using trial and error. A sequence of successful outcomes will be reinforced to develop the best recommendation or policy for a given problem.

3. Real-world machine learning use cases

Machine learning is everywhere. Here are just a few examples of machine learning you might encounter every day.

- Digital assistants: Apple Siri, Amazon Alexa, Google Assistant and other digital assistants are powered by natural language processing (NLP), a machine learning application that enables computers to process text and voice data and "understand" human language the way people do. Natural language processing also drives voice-driven applications like GPS and speech recognition (speech-to-text) software.
- Recommendations: Deep learning models drive "people also liked" and "just for you" recommendations offered by Amazon, Netflix, Spotify, and other retail, entertainment, travel, job search and news services.
- Contextual online advertising: Machine learning and deep learning models can evaluate the

content of a web page—not only the topic, but nuances like the author's opinion or attitude—and serve up advertisements tailored to the visitor's interests.

- Chatbots: Chatbots can use a combination of pattern recognition, natural language processing, and deep neural networks to interpret input text and provide suitable responses.
- Fraud detection: Machine learning regression and classification models have replaced rule-based fraud detection systems, which have a high number of false positives when flagging stolen credit card use and are rarely successful at detecting criminal use of stolen or compromised financial data.
- Cyber security: Machine learning can extract intelligence from incident reports, alerts, blog posts and more to identify potential threats, advise security analysts and accelerate response.
- Medical image analysis: The types and volume of digital medical imaging data have exploded, leading to more available information for supporting diagnoses. Convolutional Neural Networks (CNN), Recurrent Neural Networks (RNN) and other deep learning models have proven increasingly successful at extracting features and information from medical images to help support accurate diagnoses.
- Self-driving cars: Self-driving cars require a machine learning. They must continuously identify objects in the environment around the car, predict how they will change or move, and guide the car around the objects as well as toward the driver's destination. Almost every form of machine learning and deep learning algorithm mentioned above plays some role in enabling a self-driving automobile.

✎ New Words

massive	['mæsɪv]	*adj.*巨大的，大量的
prepare	[prɪ'peə]	*v.*把……准备好，为……做准备
tag	[tæg]	*vt.*加标签于，标注 *n.*标签
unlabeled	[ʌn'leɪbld]	*adj.*未标记的
properly	['prɒpəli]	*adv.*正确地；完全地；真正地
randomize	['rændəmaɪz]	*v.*使随机化
dedupe	[di'dju:p]	*v.*删除重复数值
imbalance	[ɪm'bæləns]	*n.*不平衡
bias	['baɪəs]	*n.*偏好；偏见
refine	[rɪ'faɪn]	*vt.*提炼；改善
linear	['lɪniə]	*adj.*线性的
recommendation	[ˌrekəmen'deɪʃn]	*n.*推荐；建议
pedigree	['pedɪgri:]	*n.*血统；家谱 *adj.*纯种的
proximity	[prɒk'sɪməti]	*n.*接近，邻近；接近度

cluster	['klʌstə]	v.聚集
		n.团，群，簇
record	['rekɔːd]	n.记录
prior	['praɪə]	adj.优先的；占先的
association	[ə,səʊsi'eɪʃn]	n.关系；联系；因果关系
neural	['njʊərəl]	adj.神经的
ingest	[ɪn'dʒest]	vt.接收；吸收；采集；获取
define	[dɪ'faɪn]	v.定义；阐明；限定
adjust	[ə'dʒʌst]	v.调整，调节；适应；校准
weight	[weɪt]	n.权重
spam	[spæm]	n.垃圾邮件
message	['mesɪdʒ]	n.信息；电邮
		v.给……发消息
vacuum	['vækjuːm]	n.真空；清洁
		v.用真空吸尘器清扫
supervise	['suːpəvaɪz]	v.监督；管理；指导
overfit	['əʊvəfɪt]	v.过拟合，过度拟合
unsupervise	[ʌn'suːpəvaɪz]	v.无监督；不管理
intervention	[,ɪntə'venʃn]	n.介入，干涉，干预
indicate	['ɪndɪkeɪt]	v.表明；指示
flag	[flæg]	n.标识，标记
		v.标示，标记
semi-supervised	['semɪ ʌn'suːpəvaɪzd]	adj.半监督的
reinforcement	[,riːɪn'fɔːsmənt]	n.强化，加强，增强
behavioral	[bɪ'heɪvjərəl]	adj.行为的
sample	['sɑːmpl]	n.样本
		vt.取样
trial	['traɪəl]	v.测试
		n.试验
reinforce	[,riːɪn'fɔːs]	vt.强化，增强
encounter	[ɪn'kaʊntə]	v.遭遇；偶遇
opinion	[ə'pɪnjən]	n.意见
tailor	['teɪlə]	v.专门制作；定做
chatbot	['tʃætbɒt]	n.聊天机器人
rarely	['reəli]	adv.少有地；罕见地
cyber	['saɪbə]	adj.计算机（网络）的，信息技术的
incident	['ɪnsɪdənt]	n.事件，事变；敌对行动
continuously	[kən'tɪnjuəsli]	adv.连续不断地

🐟 Phrases

training data set	训练数据集
call out	调来；召集
training subset	训练子集
regression algorithm	回归算法
logistic regression	逻辑回归
linear regression	线性回归
dependent variable	因变量，应变数，因变数
independent variable	自变量，自变数
support vector machine	支持向量机
decision tree	决策树
decision rule	决策规则
instance-based algorithm	基于实例的算法
clustering algorithm	聚类算法
association algorithm	关联算法
association rule	关联规则
data mining	数据挖掘
neural network	神经网络
input layer	输入层
hidden layer	隐藏层
output layer	输出层
deep neural network	深度神经网络
iterative process	迭代过程
semi-supervised machine learning	半监督机器学习
reinforcement machine learning	强化机器学习
digital assistant	数字助理，数字助手
fraud detection	欺诈检测
credit card	信用卡
blog post	博文，博客帖子

🐟 Abbreviations

KNN (K-Nearest Neighbor）	K 最近邻算法
GPS (Global Position System)	全球定位系统
CNN (Convolutional Neural Network)	卷积神经网络
RNN (Recurrent Neural Network)	递归神经网络

Reference Translation

机器学习

机器学习是人工智能（AI）的一个分支，专注于构建可从数据中学习并随着时间的推移而提高其准确性的应用程序，而无须编程。

在数据科学中，算法是一系列统计处理步骤。在机器学习中，对算法进行"训练"以发现大量数据中的模式和特征，以便根据新数据做出决策和预测。算法越好，随着处理的数据越多，决策和预测就越准确。

1. 机器学习的工作原理

构建机器学习应用程序（或模型）有四个基本步骤。这些通常由数据科学家与正在为其开发模型的业务专业人员紧密合作来执行。

（1）选择并准备训练数据集

训练数据是一个数据集，代表机器学习模型解决问题时将要接收的数据。在某些情况下，训练数据是标记数据——被"标记"以调出模型需要识别的特征和分类。其他数据是未标记的数据，模型将需要提取这些特征并自行分类。

无论哪种情况，都需要适当地准备训练数据——随机化、删除重复数据并检查可能影响训练的不平衡或偏差数据。它还应分为两个子集：用于训练应用程序的训练子集和用于测试与优化应用程序的评估子集。

（2）选择要在训练数据集上运行的算法

同样，算法是一组统计处理步骤。算法的类型取决于训练数据集中的数据类型（标记或未标记）和数据量以及要解决的问题的类型。

用于带有标记数据的机器学习算法的常见类型如下。

- 回归算法：线性回归和逻辑回归是用来理解数据关系的回归算法的示例。线性回归用于根据自变量的值来预测因变量的值。当因变量本质上是二元数 A 或 B 时，可以使用逻辑回归。例如，可以训练线性回归算法，根据推销员的学历或工作年限（自变量）的关系来预测推销员的年销售额（因变量）。当因变量较难分类时，另一种称为支持向量机的回归算法则很有用。
- 决策树：决策树使用分类的数据根据一组决策规则提出建议。例如，推荐投注特定马匹获胜、放置或展示的决策树可以使用有关该马匹的数据（如年龄、获胜百分比、血统），并将规则应用于这些因素以推荐一个行为或决定。
- 基于实例的算法：KNN 是基于实例的算法的一个很好的例子。它使用分类来根据数据点与其他数据点的接近程度来估计数据点成为一个或另一个组成员的可能性。

具有未标记数据的算法包括以下内容。

- 聚类算法：将聚类视为组。聚类的重点是识别相似记录的组，并根据它们所属的组对记录进行标记。无需事先了解组及其特征就可以完成此操作。聚类算法的类型包括 K-means、TwoStep 和 Kohonen 聚类。
- 关联算法：关联算法可以发现数据中的模式和关系，并识别频繁的称为关联规则的

"if-then"关系。这些规则类似于数据挖掘中使用的规则。

- 神经网络：神经网络是一种定义分层计算网络的算法，它有一个输入层（在其中接收数据）、至少一个隐藏层（在其中进行计算并得出关于输入的不同结论）和一个输出层（为每个结论分配一个概率）。深度神经网络定义了一个具有多个隐藏层的网络，每个隐藏层都在不断完善前一层的结果。

（3）训练算法以创建模型

训练算法是一个迭代过程。它涉及通过算法运行变量，将输出与其应产生的结果进行比较，调整算法中的权重和偏差（可能会产生更准确的结果），然后再次运行变量，直到大部分时间算法返回正确的结果为止。

（4）使用和改善模型

最后一步是将模型与新数据一起使用，并随着时间的推移提高模型的准确性和有效性。新数据的来源将取决于要解决的问题。例如，旨在识别垃圾邮件的机器学习模型将接收电子邮件信息，而驱动机器人吸尘器的机器学习模型将接收与移动的家具或房间中的新物体进行实际交互所产生的数据。

2. 机器学习方法

（1）监督机器学习

监督机器学习在标记的数据集上进行自我训练。也就是说，数据标记有正在构建机器学习模型的信息，甚至可以按照模型应该对数据进行分类的方式对其进行分类。例如，可以在各种标记狗图像的数据集上训练旨在识别纯种德国牧羊犬的计算机视觉模型。

监督机器学习比其他机器学习方法需要的训练数据更少，并且由于可以将模型的结果与实际标记的结果进行比较，因此使训练变得更容易。但是，标记正确的数据准备起来很昂贵，并且存在过度拟合的风险，或者会创建一个与训练数据紧密相关且有偏差的模型，以致无法正确处理新数据的变化。

（2）无监督机器学习

无监督机器学习吸收了大量的未标记数据。它使用算法来提取数据中有意义的特征，而无须人工干预。无监督学习较少涉及自动执行决策和预测，而更多的是识别人类可能遗漏的数据中的模式和关系。以垃圾邮件检测为例，人们产生的电子邮件数量超过了数据科学家可能希望对其进行标记或分类的数量，科学家穷其一生也无法完成标记和分类。一种无监督机器学习算法可以分析大量电子邮件，并发现表明垃圾邮件的特征和模式（随着时间的流逝，会不断提高对垃圾邮件的标记能力）。

（3）半监督机器学习

半监督机器学习在监督机器学习和无监督机器学习之间提供了一种很好的媒介。在训练期间，它使用较小的标记数据集来指导从较大的未标记数据集中进行分类和特征提取。半监督机器学习可以解决没有足够的标记数据（或不能负担足够的标记数据）来训练监督学习算法的问题。

（4）强化机器学习

强化机器学习是一种行为机器学习模型，类似于监督学习，但是该算法未使用示例数据进行训练。该模型通过反复试验来学习。其将加强一系列成功的结果，以针对特定问题制定最佳建议或策略。

3. 现实世界的机器学习用例

机器学习无处不在。以下只是你每天可能会遇到的一些机器学习的示例。

- 数字助理：Apple Siri、Amazon Alexa、Google Assistant 和其他数字助理都由自然语言处理（NLP）提供支持。NLP 是一种机器学习应用程序，可以让计算机处理文本和语音数据并以人类的方式"理解"人类语言。自然语言处理还推动了语音驱动的应用程序，如 GPS 和语音识别（语音到文本）软件。

- 推荐：深度学习模型推动了 Amazon、Netflix、Spotify 和其他零售、娱乐、旅行、求职与新闻服务提供的"人们也喜欢"和"只为你"的推荐。

- 语境在线广告：机器学习和深度学习模型可以评估网页的内容（不仅是主题，还可以评估作者的意见或态度之类的细微差别），并可以根据访问者的兴趣量身定制广告。

- 聊天机器人：聊天机器人可以与模式识别、自然语言处理和深度神经网络结合使用来解释输入的文本并提供适当的响应。

- 欺诈检测：机器学习回归和分类模型已经取代了基于规则的欺诈检测系统。欺诈检测系统在标记被盗信用卡被使用时会产生大量误报，并且很少能成功检测盗窃或泄露财务数据的犯罪行为。

- 网络安全：机器学习可以从事件报告、警报、博客文章等中提取情报，以识别潜在威胁，为安全分析人员提供建议并加快响应速度。

- 医学图像分析：数字医学图像数据的类型和数量呈爆炸式增长，从而提供了更多支持诊断的可用信息。卷积神经网络（CNN）、递归神经网络（RNN）和其他深度学习模型已被证明在从医学图像中提取特征和信息以帮助支持准确诊断方面越来越成功。

- 自动驾驶汽车：自动驾驶汽车需要机器学习。它们必须不断地识别汽车周围环境中的物体，预测它们将如何变化或移动，引导汽车绕过这些物体并驶向驾驶员的目的地。几乎上述的所有形式的机器学习和深度学习算法都在实现自动驾驶汽车方面发挥了一定的作用。

Text B

Deep Learning

扫码听课文

Deep learning simulates the human brain and enables systems to identify objects and perform complex tasks with increasing accuracy—all without human intervention.

1. What is deep learning

Deep learning is a subset of machine learning in which multi-layered neural networks—modeled to work like the human brain—"learn" from large amounts of data. Within each layer of the neural network, deep learning algorithms perform calculations and make predictions repeatedly, progressively "learn" and gradually improve the accuracy of the outcome over time.

In the same way that the human brain absorbs and processes information entering the body through the five senses, deep learning ingests information from multiple data sources and analyzes it in real time.

Deep learning drives many Artificial Intelligence (AI) applications and services that improve automation and perform analytical and physical tasks without human intervention. Deep learning technology lies behind everyday products and services (such as digital assistants, voice-enabled TV remotes and credit card fraud detection) as well as emerging technologies (such as self-driving cars).

2. Deep learning vs machine learning

If deep learning is a subset of machine learning, how do they differ? In the simplest terms, what sets deep learning apart from the rest of machine learning is the data it works with and how it learns.

While all machine learning can work with and learn from structured, labeled data, deep learning can also ingest and process unstructured, unlabeled data. Instead of relying on labels within the data to identify and classify objects and information, deep learning uses a multi-layered neural network to extract the features from the data and get better and better at identifying and classifying data on its own.

For example, the voice-to-text applications of a decade ago (which users had to train by speaking scores of words to the application and, in the process, label their own voice data) are examples of machine learning. Today's voice recognition applications (including Apple Siri, Amazon Alexa, and Google Assistant), which can recognize anyone's voice commands without a specific training session, are examples of deep learning.

In more technical terms, while all machine learning models are capable of supervised learning (requiring human intervention), deep learning models are also capable of unsupervised learning. They can detect previously undetected features or patterns in data that aren't labeled, with the barest minimum of human supervision. Deep learning models are also capable of reinforcement learning— a more advanced unsupervised learning process in which the model "learns" to become more accurate based on positive feedback from previous calculations.

3. How deep learning works

Deep learning neural networks (called deep neural networks) are modeled on the way scientists believe the human brain works. They process data repeatedly and refine the analysis and results gradually to accurately recognize, classify and describe objects within the data.

Deep neural networks consist of multiple layers of interconnected nodes, each of which uses a progressively more complex deep learning algorithm to extract and identify features and patterns in the data. They then calculate the likelihood that the object or information can be classified or identified in one or more ways.

The input and output layers of a deep neural network are called visible layers. The input layer is where the deep learning model ingests the data for processing, and the output layer is where the final identification, classification or description is calculated.

In between the input and output layers are hidden layers where the calculations of each previous layer are weighted and refined by progressively more complex algorithms. This movement of calculations through the network is called forward propagation.

Another process called back propagation identifies errors in calculated predictions, assigns them weights and biases, and pushes them back to previous layers to train or refine the model. Together, forward propagation and back propagation allow the network to make predictions about the identity or class of the object while learning from inconsistencies in the outcomes. The result is a system that learns as it works and gets more efficient and accurate over time when processing large amounts of data.

The above describes the simplest type of deep neural network in the simplest terms. In practice, deep learning algorithms are incredibly complex. And many complex deep learning methods and models have been developed to solve certain types of problems, including the following examples.

- Convolutional Neural Networks (CNN), used primarily in computer vision applications, can detect features and patterns within a complex image and, ultimately, recognize specific objects within the image. In 2015, a CNN bested a human in an object recognition challenge for the first time.

- Recurrent Neural Networks (RNN) are used for deep learning models in which features and patterns change over time. Instead of ingesting and outputting data snapshots, RNN ingest and output sequences of data. RNN drive emerging applications such as speech recognition and driverless cars.

One of the best ways to improve your understanding of deep learning networks is through video that illustrates the way data moves through various deep learning models.

4. Deep learning applications

Real-world deep learning applications are a part of our daily lives, but in most cases, they are so well-integrated into products and services that users are unaware of the complex data processing that is taking place in the background. The following are some of these examples.

(1) Law enforcement

Deep learning algorithms can analyze and learn from transactional data to identify dangerous patterns that indicate possible fraudulent or criminal activity. Speech recognition, computer vision and other deep learning applications can improve the efficiency and effectiveness of investigative analysis by extracting patterns and evidence from sound and video recordings, images and documents, which helps law enforcement analyze large amounts of data more quickly and accurately.

(2) Financial services

Financial institutions regularly use predictive analytics to drive algorithmic trading of stocks, assess business risks for loan approvals, detect fraud, and help manage credit and investment portfolios for clients.

(3) Customer service

Many organizations incorporate deep learning technology into their customer service processes. Chatbots, which are used in a variety of applications, services and customer service portals, are a straight forward form of AI. Traditional chatbots use natural language and even visual recognition, commonly found in call center-like menus. However, more sophisticated chatbot solutions attempt to determine through learning if there are multiple responses to ambiguous questions. Based on the responses it receives, the chatbot then tries to answer these questions directly or streamline the dialogue transition to a human user.

Virtual assistants like Apple Siri, Amazon Alexa or Google Assistant add a third dimension to the chatbot concept by combining deep learning capabilities with the underlying technology. These data science innovations allow for speech recognition and customized responses, resulting in a personalized experience for the users.

(4) Healthcare

The healthcare industry has benefited greatly from deep learning ever since the digitization of hospital records and images. Image recognition applications can support medical imaging specialists and radiologists and help them analyze and assess more images in less time.

✍ New Words

increase	[ɪnˈkriːs]	v.增加，增长
	[ˈɪnkriːs]	n.增加，增长
layer	[ˈleɪə]	n.层，层次
		v.分层
repeatedly	[rɪˈpiːtɪdli]	adv.反复地，重复地
progressively	[prəˈgresɪvli]	adv.日益增加地；逐步地
improve	[ɪmˈpruːv]	v.改进，提高
absorb	[əbˈzɔːb]	v.吸收；理解，掌握
Sense	[sens]	n.感觉官能（即视、听、嗅、味、触五觉）；感觉
		v.感觉到；意识到
voice	[vɔɪs]	n.语音
capable	[ˈkeɪpəbl]	adj.有能力的；胜任的
undetected	[ˌʌndɪˈtektɪd]	adj.未被察觉的，未被发现的
positive	[ˈpɒzətɪv]	adj.正面的
likelihood	[ˈlaɪklihʊd]	n.可能性
visible	[ˈvɪzəbl]	adj.可见的
description	[dɪˈskrɪpʃn]	n.描述，说明；种类，性质
inconsistency	[ˌɪnkənˈsɪstənsi]	n.不一致，不协调；前后矛盾
incredibly	[ɪnˈkredəbli]	adv.难以置信地，很，极为
snapshot	[ˈsnæpʃɒt]	n.快照
unaware	[ˌʌnəˈweə]	adj.不知道的；未察觉到的；不注意的
enforcement	[ɪnˈfɔːsmənt]	n.实施，执行
investigative	[ɪnˈvestɪgətɪv]	adj.调查性质的；研究的
evidence	[ˈevɪdəns]	n.证据
		v.证明
stock	[stɒk]	n.股份，股票；库存

menu	['menju:]	n.菜单
sophisticate	[sə'fɪstɪkeɪt]	adj.有经验的，老于世故的
ambiguous	[æm'bɪgjuəs]	adj.模棱两可的；不明确的
receive	[rɪ'si:v]	v.收到，得到
streamline	['stri:mlaɪn]	vt.流畅；使简单化，使现代化
dialogue	['daɪəlɒg]	n.对话
innovation	[ˌɪnə'veɪʃn]	n.改革，创新；新观念；新发明
digitization	[ˌdɪdʒɪtaɪ'zeɪʃən]	n.数字化
specialist	['speʃəlɪst]	n.专家；专科医生

✎ Phrases

multi-layered neural network	多层神经网络
real time	实时
interconnected node	互联的节点
visible layer	可见层
forward propagation	前向传播
back propagation	反向传播
data snapshot	数据快照
driverless car	无人驾驶汽车
financial institution	金融机构
loan approval	贷款批准
investment portfolios	投资组合
a variety of	各种各样的
customized response	客户化响应，自定义响应

Reference Translation

深度学习

深度学习可以模拟人的大脑，使系统能够以越来越高的精度识别对象和执行复杂任务——所有这些都无须人工干预。

1. 什么是深度学习

深度学习是机器学习的一个子集，其中，多层神经网络（像人的大脑一样工作的模型）从大量数据中"学习"。在神经网络的每一层中，深度学习算法会反复执行计算和预测，逐步"学习"并随着时间的推移逐渐提高结果的准确性。

就像人脑通过五种感官吸收并处理进入人体的信息一样，深度学习从多个数据源中采集信息并进行实时分析。

深度学习推动了许多人工智能（AI）应用程序和服务，这些应用程序和服务可以提高自动化程度，并在无须人工干预的情况下执行分析和物理任务。深度学习技术支持日常产品和服务（如数字助理、启用语音的电视遥控器和信用卡欺诈检测）以及新兴技术（如自动驾驶汽车）。

2. 深度学习与机器学习

如果深度学习是机器学习的子集，它们有何不同？用最简单的术语来说，将深度学习与其他机器学习区分开来的是它所使用的数据及其学习方式。

虽然所有机器学习都可以使用结构化、标记的数据并从中学习，但深度学习还可以采集和处理非结构化、未标记的数据。深度学习不是依靠数据中的标签来识别和分类对象与信息，而是使用多层神经网络从数据中提取特征，并越来越擅长自行识别和分类数据。

例如，十年前的语音转文本应用程序（用户必须通过向应用程序说出大量单词来进行训练，并在此过程中标记自己的语音数据）是机器学习的示例。如今的语音识别应用程序（包括 Apple Siri、Amazon Alexa 和 Google Assistant）可以在无须特定培训的情况下识别任何人的语音命令，这些都是深度学习的示例。

用更技术的术语来说，尽管所有机器学习模型都可以进行监督学习（需要人工干预），但深度学习模型也可以进行无监督学习。它们可以在没有人为监督的情况下，检测未标记数据中以前未检测到的特征或模式。深度学习模型还能够进行强化学习，这是一种更高级的无监督学习过程，在该过程中，模型会根据先前计算的正反馈"学习"，变得更准确。

3. 深度学习的工作原理

深度学习神经网络（称为深度神经网络）以科学家认为人脑工作的方式为模型。它们反复处理数据，逐步完善分析和结果，以准确识别、分类和描述数据中的对象。

深度神经网络由多层相互连接的节点组成，每个节点都使用越来越复杂的深度学习算法来提取和识别数据中的特征与模式。然后，它们计算可以通过一种或多种方式对对象或信息进行分类或识别的可能性。

深度神经网络的输入层和输出层称为可见层。输入层是深度学习模型提取数据进行处理的地方，输出层是进行最终标识、分类或描述的地方。

在输入层和输出层之间是隐藏层，其中每个前一层的计算将通过越来越复杂的算法进行加权和优化。这种通过网络进行的计算运动称为前向传播。

另一个称为反向传播的过程可以识别计算出的预测中的错误，为它们分配权重和偏差，然后将它们推回到上一层从而训练或完善模型。前向传播和反向传播一起使网络在可以从结果的不一致中学习的同时，对对象的身份或类别做出预测。结果是系统可以边工作边学习，并且在处理大量数据时会随着时间的推移变得更加高效和准确。

上面用最简单的术语描述了最简单的深度神经网络类型。在实践中，深度学习算法非常复杂。为了解决某些类型的问题，已经开发了许多复杂的深度学习方法和模型，包括以下示例。

- 卷积神经网络（CNN）主要用于计算机视觉应用程序，可以检测复杂图像中的特征和模式，并最终识别图像中的特定对象。2015 年，CNN 首次在对象识别挑战中击败了人类。
- 递归神经网络（RNN）用于深度学习模型，其中特征和模式随时间变化。RNN 不是采集和输出数据快照，而是采集和输出数据序列。RNN 推动了新兴应用，如语音识别和无人驾驶汽车。

提高对深度学习网络理解的最佳方法之一是通过视频来说明数据在各种深度学习模型中的移动方式。

4. 深度学习应用

现实世界中的深度学习应用程序是我们日常生活的一部分，但是在大多数情况下，它们已经很好地集成到了产品和服务中，以至于用户不知道后台在进行的复杂数据处理。以下是其中一些示例。

（1）执法

深度学习算法可以分析交易数据并从中学习，以识别出可能存在欺诈或犯罪活动的危险模式。语音识别、计算机视觉和其他深度学习应用程序可以通过从声音和视频记录、图像和文档中提取模式与证据来提高调查分析的效率和有效性，从而帮助执法机构更快、更准确地分析大量数据。

（2）金融服务

金融机构定期使用预测分析来推动股票的算法交易、评估贷款批准的业务风险、检测欺诈并帮助管理客户的信贷和投资组合。

（3）客户服务

许多组织将深度学习技术整合到它们的客户服务流程中。聊天机器人用于各种应用程序、服务和客户服务门户，是一种直接的人工智能形式。传统的聊天机器人使用自然语言甚至是视觉识别，这通常在类似呼叫中心的菜单中很常见。但是，更复杂的聊天机器人解决方案试图通过学习来确定是否对模棱两可的问题有多个响应。然后，聊天机器人将根据收到的响应尝试直接回答这些问题，或者顺畅地过渡到与人类用户的对话。

Apple Siri、Amazon Alexa 或 Google Assistant 等虚拟助理通过将深度学习功能与基础技术相结合，为聊天机器人概念增加了新的内容。这些数据科学创新可实现语音识别和自定义响应，从而为用户带来个性化的体验。

（4）医疗保健

自从医院记录和图像数字化以来，医疗保健行业已从深度学习中受益匪浅。图像识别应用程序可以为医学成像专家和放射科医生提供支持，并帮助他们在更短的时间内分析和评估更多的图像。

Exercises

[Ex. 1] Answer the following questions according to Text A.

1. What is machine learning?

2. What are algorithms "trained" to do in machine learning?

3. What is training data?

4. What are the common types of machine learning algorithms with labeled data?

5. What do the algorithms with unlabeled data include?

6. Why does supervised machine learning make training easier?

7. What is unsupervised learning less about? And what is it more about?

8. What is reinforcement machine learning?

9. What can machine learning and deep learning models do in contextual online advertising?

10. What can machine learning do in cyber security?

[Ex. 2] Fill in the following blanks according to Text B.

1. Deep learning is a subset of _____ in which multi-layered neural networks—modeled to work like the human brain—"learn" from _____.

2. While all machine learning can work with and learn from _____, _____ data, deep learning can also ingest and process _____, _____ data.

3. Instead of relying on _____ to identify and classify objects and information, deep learning uses _____ to extract the features from the data and get better and better at _____ and _____ data on its own.

4. In more technical terms, while all machine learning models are capable of _____, deep learning models are also capable of _____.

5. Deep neural networks consist of _____ of interconnected nodes, each of which uses a progressively more complex _____ to extract and identify _____ in the data.

6. The input layer is where the deep learning model _____ for processing, and the output layer is where _____, _____ or _____ is calculated.

7. Together, _____ and _____ allow the network to make predictions about the identity or class of the object while learning from _____ in the outcomes.

8. Convolutional Neural Networks (CNN), used primarily in _____, can detect _____ within a complex image and, ultimately, recognize _____ within the image.

9. Speech recognition, computer vision and other deep learning applications can improve _____ and _____ of investigative analysis by extracting patterns and evidence from _____, images and _____, which helps law enforcement analyze large amounts of data _____.

10. The healthcare industry has benefited greatly from _____ ever since the digitization of _____ and images. Image recognition applications can support _____ and radiologists and help them _____ more images in less time.

[Ex. 3] Translate the following terms or phrases from English into Chinese.

1.	adjust	1.	_____
2.	chatbot	2.	_____
3.	cluster	3.	_____
4.	flag	4.	_____
5.	ingest	5.	_____
6.	association algorithm	6.	_____

7. clustering algorithm	7. _____
8. data mining	8. _____
9. decision tree	9. _____
10. deep neural network	10. _____

[Ex. 4] Translate the following terms or phrases from Chinese into English.

1. *adj.*线性的	1. _____
2. *adj.*神经的	2. _____
3. *n.*推荐；建议	3. _____
4. *vt.*强化，增强	4. _____
5. *v.*监督；管理；指导	5. _____
6. 数字助理，数字助手	6. _____
7. 自变量，自变数	7. _____
8. 迭代过程	8. _____
9. 线性回归	9. _____
10. 半监督机器学习	10. _____

[Ex. 5] Translate the following passage into Chinese.

Machine Learning Models

The prime goal of a machine learning algorithm is to generalize from its experience and then use the acquired knowledge in order to solve new, previously unknown and unseen tasks. Just like humans, actually.

There are 9 different types of machine learning algorithms, each serving a different purpose and using distinct tools. Let's take a look at them.

- Supervised learning: Here machine learning algorithm is presented with both input and the desired output data before it starts learning, the same human procedure we call "process learning".

- Unsupervised learning: As opposed to the previous type, this machine learning algorithm is only presented with input data and has to structure data on its own.

- Semi-supervised learning: It is a machine learning method that uses combined parts of data from supervised learning and parts from unsupervised learning types. It can produce heightened learning accuracy quality.

- Reinforcement learning: It is the machine learning type you are familiar with quite a lot, as it teaches a computer algorithm how to play as a human opponent.

- Self learning: It is a machine learning type that uses no external rewards and no external teacher advises.

- Feature learning: It replaces manual feature engineering and enables a machine learning algorithm to both learn the features and utilize them to perform a specific task.

- Sparse dictionary learning: It is a machine learning method whose goal is to find a sparse representation of the input data in the form of a linear combination of basic elements, which are called atoms and compose a dictionary.
- Anomaly detection: It can recognition of untypical events, items or bursts in activity that are significantly different from the majority of the data.
- Association rules: It is a rule-based machine learning method for identifying relations and strong rules among variables in large databases.

Unit 7

Text A

扫码听课文

Top 10 Machine Learning Algorithms (I)

According to a recent study, machine learning algorithms are expected to replace 25% of the jobs across the world in the next 10 years. With the rapid growth of big data and the introduction of programming tools like Python and R, machine learning is gaining mainstream presence for data scientists. Machine learning applications are highly automated and self-modifying. They continue to improve over time with minimal human intervention as they learn with more data. For instance, Netflix's recommendation algorithm learns more about the likes and dislikes of a viewer based on the shows every viewer watches. Specialized machine learning algorithms have been developed to solve complex problems perfectly.

1. Naive Bayes classifier algorithm

It is difficult and practically impossible to classify a web page, a document, an email or any other lengthy text notes manually. This is where naive Bayes classifier algorithm comes to the rescue. A classifier is a function that allocates a population's element value from one of the available categories. For instance, spam filtering is a popular application of naive Bayes algorithm. Spam filter, here, is a classifier that assigns a label "Spam" or "Not Spam" to all the emails.

Naive Bayes classifier is among the most popular learning method grouped by similarities. It works on the popular Bayes probability theorem to build machine learning models, particularly for disease prediction and document classification. It is a simple classification of words based on Bayes probability theorem for subjective analysis of content.

(1) When to use naive Bayes classifier algorithm

- If you have a moderate or large training data set.
- If the instances have several attributes.
- Given the classification parameter, attributes which describe the instances should be conditionally independent.

(2) Applications of naive Bayes classifier

- Sentiment analysis: It is used at Facebook to analyse status updates expressing positive or negative emotions.
- Document categorization: Google uses document classification to index documents and find relevancy scores i.e. the PageRank.
- Naive Bayes algorithm is also used for classifying news articles about technology,

entertainment, sports, politics, etc.

- Email spam filtering: Google Mail uses naive Bayes algorithm to classify your emails as spam or not spam.

(3) Advantages of naive Bayes classifier algorithm

- Naive Bayes classifier algorithm performs well when the input variables are categorical.
- A naive Bayes classifier converges faster and requires relatively little training data than other discriminative models like logistic regression when the naive Bayes conditional independence assumption holds.
- With naive Bayes classifier algorithm, it is easier to predict class of the test data set. A good bet for multi class predictions as well.
- Though it requires conditional independence assumption, naive Bayes classifier has presented good performance in various application domains.

2. K-means clustering algorithm

K-means is a popularly used unsupervised machine learning algorithm for cluster analysis. K-means is a non-deterministic and iterative method. The algorithm operates on a given data set through pre-defined number of clusters k. The output of K-means algorithm is k clusters with input data partitioned among the clusters.

For instance, let's consider K-means clustering for Wikipedia search results. The search term "Jaguar" on Wikipedia will return all pages containing the word Jaguar. K-means clustering algorithm can be applied to group the web pages that talk about similar concepts.

(1) Advantages of K-means clustering algorithm

- In case of globular clusters, K-means produces tighter clusters than hierarchical clustering.
- Given a smaller value of k, K-means clustering computes faster than hierarchical clustering for large number of variables.

(2) Applications of K-means clustering

K-means clustering algorithm is used by most of the search engines like Yahoo, Google to cluster web pages by similarity and identify the "relevancy rate" of search results. This helps search engines reduce the computational time for the users.

3. Support vector machine algorithm

Support Vector Machine (SVM) is a supervised machine learning algorithm for classification or regression problems where the data set teaches SVM about the classes so that SVM can classify any new data. It works by classifying the data into different classes by finding a line (hyperplane) which separates the training data set into classes. As there are many such linear hyperplanes, SVM algorithm tries to maximize the distance between the various classes that are involved and this is referred to as margin maximization.

(1) Categories of SVM

- Linear SVM: In linear SVM the training data i.e. classifiers are separated by a hyperplane.
- Non-Linear SVM: In non-linear SVM it is not possible to separate the training data using a hyperplane. For example, the training data for face detection consists of group of images that

are faces and another group of images that are not faces (in other words all other images in the world except faces). Under such conditions, the training data is too complex that it is impossible to find a representation for every feature vector. Separating the set of faces linearly from the set of non-face is a complex task.

(2) Advantages of SVM

- It offers best classification performance (accuracy) on the training data.
- It renders more efficiency for correct classification of the future data.
- It does not make any strong assumptions on data, which is the best thing about SVM.
- It does not overfit the data.

(3) Applications of SVM

SVM is commonly used for stock market forecasting by various financial institutions. For instance, it can be used to compare the relative performance of the stocks when compared to performance of other stocks in the same sector. The relative comparison of stocks helps manage investment making decisions based on the classifications made by the SVM learning algorithm.

4. Apriori algorithm

Apriori algorithm is an unsupervised machine learning algorithm that generates association rules from a given data set. Association rule implies that if an item A occurs, then item B also occurs with a certain probability. Most of the association rules generated are in the if-then format. For example, if people buy an iPad then they also buy an iPad case to protect it. For the algorithm to derive such conclusions, it first observes the number of people who bought an iPad case while purchasing an iPad. This way a ratio is derived like out of the 100 people who purchased an iPad, 85 people also purchased an iPad case.

Principles on which Apriori algorithm works are as follows.

- If an item set occurs frequently then all the subsets of the item set also occur frequently.
- If an item set occurs infrequently then all the supersets of the item set have infrequent occurrence.

(1) Advantages of Apriori algorithm

- It is easy to implement and can be parallelized easily.
- Apriori algorithm implementation makes use of large item set properties.

(2) Applications of Apriori algorithm

- Detecting adverse drug reactions: Apriori algorithm is used for association analysis on healthcare data like the drugs taken by patients, characteristics of each patient, ill-effects patients experience, initial diagnosis, etc. This analysis produces association rules that help identify the combination of patient characteristics and medications that lead to adverse side effects of the drugs.
- Market basket analysis: Many e-commerce giants like Amazon use Apriori to draw data insights on which products are likely to be purchased together and which are most responsive to promotion. For example, a retailer might use Apriori to predict that people who buy sugar and flour are likely to buy eggs to bake a cake.

- Auto-complete applications: Google auto-complete is another popular application of Apriori algorithm wherein when the user types a word, the search engine looks for other associated words that people usually type after a specific word.

5. Linear regression algorithm

Linear regression algorithm shows the relationship between 2 variables and how the change in one variable impacts the other. The algorithm shows the impact on the dependent variable on changing the independent variable. The independent variables are referred to as explanatory variables, as they explain the factors the impact the dependent variable. Dependent variable is often referred to as the factor of interest or predictor.

(1) Advantages of linear regression algorithm

- It is one of the machine learning algorithms with interpretability.
- It is easy of use as it requires minimal tuning.
- It is the mostly widely used machine learning technique that runs fast.

(2) Applications of linear regression

- Estimating sales: Linear regression finds great use in business, for example sales forecasting based on the trends. If a company observes steady increase in sales every month, a linear regression analysis of the monthly sales data helps the company forecast sales in upcoming months.
- Risk assessment: Linear regression helps assess risk involved in insurance or financial domain. A health insurance company can do a linear regression analysis on the number of claims per customer against age. This analysis helps insurance companies find that older customers tend to make more insurance claims. Such analysis results play a vital role in important business decisions and are made to account for risk.

New Words

expect	[ɪkˈspekt]	v.预期；盼望；料想
rapid	[ˈræpɪd]	adj.快速的；瞬间的
growth	[grəʊθ]	n.发展，增加，增长
introduction	[ˌɪntrəˈdʌkʃn]	n.介绍；引进；推出；推行
mainstream	[ˈmeɪnstriːm]	n.主流；主要倾向，主要趋势
classify	[ˈklæsɪfaɪ]	v.将……分类
Rescue	[ˈreskjuː]	v.营救，救助 n.营救（行动）
subjective	[səbˈdʒektɪv]	adj.主观的；个人的
moderate	[ˈmɒdərət]	adj.中等的
conditionally	[kənˈdɪʃənəli]	adv.有条件地
relevancy	[ˈreləvənsi]	n.关联；恰当；关联事物
entertainment	[ˌentəˈteɪnmənt]	n.娱乐节目，娱乐活动

converge	[kən'vɜ:dʒ]	v.汇集，聚集；收敛
discriminative	[dɪs'krɪmɪnətɪv]	adj.有判别力的
non-deterministic	[nɒn dɪˌtɜ:mɪ'nɪstɪk]	adj.非确定性的，不确定的
pre-defined	[pri: dɪ'faɪnd]	adj.预定义的
Partition	[pɑ:'tɪʃn]	vt.分段，分开，隔开 n.隔离物
similarity	[ˌsɪmə'lærəti]	n.相似度
regression	[rɪ'greʃn]	n.回归
hyperplane	['haɪpəˌpleɪn]	n.超平面
non-linear	[nɒn 'lɪniə]	adj.非线性的
condition	[kən'dɪʃn]	n.状况；环境；条件
impossible	[ɪm'pɒsəbl]	adj.不可能的
render	['rendə]	v.造成；给予；表达
investment	[ɪn'vestmənt]	n.投资
occurrence	[ə'kʌrəns]	n.发生，出现
superset	['sju:pəset]	n.超集，扩展集，父集
infrequent	[ɪn'fri:kwənt]	adj.稀少的；罕见的
parallelize	['pærəleˌlaɪz]	vt.使并行，使平行
promotion	[prə'məʊʃn]	n.（商品等的）促销，推广
explanatory	[ɪk'splænətri]	adj.解释性的，说明性的
predictor	[prɪ'dɪktə]	n.预言者，预报器
interpretable	[ɪn'tɜ:prətəbl]	adj.能说明的，能解释的，可判断的
tuning	['tju:nɪŋ]	n.调整

Phrases

programming tool	程序设计工具，编程工具
naive Bayes classifier algorithm	朴素贝叶斯分类器算法
Bayes probability theorem	贝叶斯概率定理
sentiment analysis	情绪分析，情感分析
negative emotion	消极情绪，负面情绪
iterative method	迭代法
K-means clustering	K 均值聚类
globular cluster	球形聚类
hierarchical clustering	层次聚类
search engine	搜索引擎
relevancy rate	相关率，关联率

feature vector	特征向量
stock market	股票市场，股票行情
item set	项目集
adverse drug reaction	药物不良反应
vital role	重要作用，重要角色

 Abbreviations

SVM (Support Vector Machine)	支持向量机

Reference Translation

十大机器学习算法（I）

根据最近的一项研究，机器学习算法有望在未来 10 年内取代全球 25％的工作。随着大数据的快速增长以及 Python 和 R 等编程工具的推出，机器学习正在数据科学家的研究中占据主流地位。机器学习应用程序是高度自动化和可自我修改的。随着时间的推移，由于使用更多的数据进行学习，它们以最小的人工干预不断改进。例如，Netflix 的推荐算法会根据每个观众所看的节目更多地了解其好恶。现在已经开发出专门的机器学习算法来完美地解决复杂问题。

1. 朴素贝叶斯分类器算法

手动分类网页、文档、电子邮件或任何其他冗长的文本注释非常困难，实际上几乎是不可能的。这就是朴素贝叶斯分类器算法的用武之地。分类器是一个从类别中分配总体元素值的函数。例如，垃圾邮件过滤是朴素贝叶斯算法的一种流行应用。这里的垃圾邮件过滤器是一个分类器，为所有电子邮件分配"垃圾邮件"或"非垃圾邮件"标签。

朴素贝叶斯分类器是按相似性分组的最受欢迎的学习方法之一。它基于流行的贝叶斯概率定理来构建机器学习模型，尤其是用于疾病预测和文档分类的机器学习模型。它是基于贝叶斯概率定理的单词的简单分类，用于内容的主观分析。

（1）何时使用朴素贝叶斯分类器算法
- 如果你具有中等或较大的训练数据集。
- 实例是否具有有多个属性。
- 给定分类参数，描述实例的属性应该是条件独立的。

（2）朴素贝叶斯分类器的应用
- 情绪分析：在 Facebook 上用于分析表达积极情绪或消极情绪的状态更新。
- 文档分类：Google 使用文档分类为文档建立索引并找到相关性得分，即 PageRank。
- 朴素贝叶斯算法也用于对有关技术、娱乐、体育、政治等方面的新闻文章进行分类。
- 电子邮件垃圾邮件过滤：Google Mail 使用朴素贝叶斯算法将你的电子邮件分类为垃圾邮件或非垃圾邮件。

（3）朴素贝叶斯分类器算法的优点

- 当输入变量为分类变量时，朴素贝叶斯分类器算法表现良好。
- 当朴素贝叶斯条件独立性假设成立时，朴素贝叶斯分类器比其他判别模型（如逻辑回归）的收敛速度更快，并且所需的训练数据要少。
- 使用朴素贝叶斯分类器算法，可以更轻松地预测测试数据集的类别。对于多类别预测也是一个不错的选择。
- 尽管需要条件独立性假设，但朴素贝叶斯分类器在各种应用领域中都表现出良好的性能。

2. K 均值聚类算法

K 均值聚类算法是一种广泛用于聚类分析的无监督机器学习算法。K 均值聚类算法是一种不确定的迭代方法。该算法通过预先定义的簇数 k 对给定的数据集进行操作。K 均值聚类算法的输出是 k 个簇，输入数据在簇之间分配。

例如，我们在 Wikipedia 中用 K 均值聚类搜索。在 Wikipedia 中输入搜索词"Jaguar"，将返回含有 Jaguar 的所有页面。K 均值聚类算法可以应用于对讨论类似概念的网页进行分组。

（1）K 均值聚类算法的优点

- 对于球形聚类，它产生的聚类比层次聚类更紧密。
- 给定较小的 k 值，对于大量变量，它的计算速度比分层聚类更快。

（2）K 均值聚类的应用

K 均值聚类算法被 Yahoo、Google 等大多数搜索引擎所采用，通过相似度对网页进行聚类并确定搜索结果的"相关率"。这有助于搜索引擎减少用户的计算时间。

3. 支持向量机算法

支持向量机（SVM）是一种用于分类或回归问题的监督机器学习算法，其中数据集向 SVM 教授有关类的信息，以便 SVM 可以对任何新数据进行分类。它的工作原理是通过找到将训练数据集划分为不同类的一条线（超平面），将数据分为不同的类。由于存在许多此类线性超平面，因此 SVM 算法试图使其涉及的各个类别之间的距离最大化，这被称为余量最大化。

（1）SVM 的类别

- 线性 SVM：在线性 SVM 中，训练数据（即分类器）由超平面分隔。
- 非线性 SVM：在非线性 SVM 中，无法使用超平面分离训练数据。例如，用于面部检测的训练数据包括作为面部的图像组和不是面部的另一组图像（换句话说，世界上除面部以外的所有其他图像）。在这样的条件下，训练数据太复杂以至于不可能为每个特征向量找到一个表示形式。将面部集与非面部集线性分离是一项复杂的任务。

（2）SVM 的优点

- 它在训练数据上提供最佳的分类性能（准确性）。
- 它为将来数据的正确分类提供了更高的效率。
- 它没有对数据做任何强有力的假设，这对 SVM 来说是最好的。
- 它不会过度拟合数据。

（3）支持向量机的应用

支持向量机通常用于各种金融机构的股票市场预测。例如，当与同一行业中其他股票的表现进行比较时，它可以用来比较股票的相对表现。股票的相对比较有助于管理基于 SVM

学习算法分类的投资决策。

4. 先验算法

先验算法是一种无监督机器学习算法，可根据给定的数据集生成关联规则。关联规则意味着如果出现项目 A，则项目 B 也将以一定概率出现。生成的大多数关联规则均为 if-then 格式。例如，如果人们购买了 iPad，那么他们还会购买 iPad 保护套来保护它。为了使算法得出这样的结论，它首先观察了在购买 iPad 的同时也购买了 iPad 保护套的人数。这样一来，就可以得出比例，例如在购买 iPad 的 100 人中，有 85 人也购买了 iPad 保护套。

Apriori 算法的工作原理如下。

- 如果某个项目集频繁出现，则该项目集的所有子集也会频繁出现。
- 如果一个项目集很少出现，则该项目集的所有超集都很少出现。

（1）先验算法的优点

- 易于实现，并且可以很容易地并行化。
- 先验算法的实现利用了大型项目集属性。

（2）先验算法的应用

- 检测药物不良反应：先验算法用于对医疗数据进行关联分析，如患者服用的药物、每位患者的特征、患者经历的不良反应、初步诊断等。此分析产生的关联规则有助于确定导致药物副作用的患者特征和药物的组合。
- 购物篮分析：许多像亚马逊这样的电子商务巨头都使用先验算法来获取数据见解，以了解哪些产品可能一起购买以及哪些产品对促销反应最灵敏。例如，零售商可能使用先验算法来预测购买糖和面粉的人可能会购买鸡蛋以烘烤蛋糕。
- 自动完成应用程序：Google 自动完成是先验算法的另一种流行的应用程序，其中，当用户键入一个单词时，搜索引擎会寻找人们通常在特定单词之后键入的其他关联单词。

5. 线性回归算法

线性回归算法显示了两个变量之间的关系以及一个变量的变化是如何影响另一个变量。该算法显示了自变量变化对因变量的影响。自变量被称为解释变量，因为它们解释了影响因变量的因素。因变量通常被称为关注因子或预测因子。

（1）线性回归算法的优点

- 它是具有可解释性的机器学习算法之一。
- 它易于使用，因为它需要最少的调整。
- 它是运行速度最快、使用最广泛的机器学习技术。

（2）线性回归的应用

- 估计销售额。线性回归在业务中非常有用，如基于趋势的销售预测。如果公司观察到每月销售量稳定增长，则对月度销售数据进行线性回归分析有助于公司预测未来几个月的销售量。
- 风险评估。线性回归有助于评估涉及保险或金融领域的风险。健康保险公司可以对每个客户的索赔数量与年龄进行线性回归分析。该分析可以帮助保险公司发现年龄较大的客户倾向于提出更多的保险索赔。此类分析结果在重要的业务决策中起着至关重要的作用，并被用于考虑风险。

Text B

扫码听课文

Top 10 Machine Learning Algorithms (II)

6. Logistic regression algorithm

The name of this algorithm could be a little confusing in the sense that logistic regression algorithm is for classification tasks and not regression problems. The name "regression" here implies that a linear model is fit into the feature space. This algorithm applies a logistic function to a linear combination of features to predict the outcome of a categorical dependent variable based on predictor variables. It helps estimate the probability of falling into a specific level of the categorical dependent variable based on the given predictor variables.

(1) Types of logistic regression

- Binary logistic regression: The most commonly used logistic regression when the categorical response has 2 possible outcomes i.e. either yes or not. Example: Predicting whether a student will pass or fail an exam, predicting whether a student will have low or high blood pressure, predicting whether a tumour is cancerous or not.
- Multi-nominal logistic regression: Categorical response has 3 or more possible outcomes with no ordering. Example: Predicting what kind of search engine (Yahoo, Bing, Google and MSN) is used by majority of US citizens.
- Ordinal logistic regression: Categorical response has 3 or more possible outcomes with natural ordering. Example: How a customer rates the service and quality of food at a restaurant based on a scale of 1 to 10.

(2) Advantages of logistic regression

- It is easier to inspect and less complex.
- It is a robust algorithm as the independent variables need not have equal variance or normal distribution.
- These algorithms do not assume a linear relationship between the dependent and independent variables and hence can also handle non-linear effects.
- It controls confounding and tests interaction.

(3) Applications of logistic regression

- It is applied in the field of epidemiology to identify risk factors for diseases and plan accordingly for preventive measures.
- It is used to classify a set of words as nouns, pronouns, verbs, adjectives.
- It is used in weather forecasting to predict the probability of rain.
- It is used in credit scoring systems for risk management to predict the defaulting of an account.

7. Decision tree algorithm

A decision tree is a graphical representation that makes use of branching methodology to exemplify all possible outcomes of a decision based on certain conditions. In a decision tree, the internal node represents a test on the attribute, each branch of the tree represents the outcome of the test and the leaf node represents a particular class label. That is the decision is made after computing all of the attributes. The classification rules are represented through the path from root to the leaf node.

There are two types of decision trees: classification trees and regression trees. They can also be classified into two types based on the type of target variable: continuous variable decision trees and binary variable decision trees. It is the target variable that helps decide what kind of decision tree would be required for a particular problem.

(1) Advantages of decision tree algorithms

- Decision trees are very instinctual and can be explained to anyone with ease.
- When using decision tree algorithms, data type is not a constraint as they can handle both categorical and numerical variables.
- Decision tree algorithms do not require making any assumption on the linearity in the data and hence can be used in circumstances where the parameters are non-linearly related.
- These algorithms are useful in data exploration.
- Decision trees help save data preparation time, as they are not sensitive to missing values and outliers.

(2) Applications of decision tree algorithm

- Decision trees are among the popular machine learning algorithms that find great use in finance for option pricing.
- Remote sensing is an application area for pattern recognition based on decision trees.
- Decision tree algorithms are used by banks to classify loan applicants by their probability of defaulting payments.

8. Random forest algorithm

Random forest is the go-to algorithm that uses a bagging approach to create a bunch of decision trees with random subset of the data. A model is trained several times on random sample of the dataset to achieve good prediction performance from the random forest algorithm. In this ensemble learning method, the output of all the decision trees in the random forest is combined to make the final prediction. The final prediction of the random forest algorithm is derived by polling the results of each decision tree or just by going with a prediction that appears the most times in the decision trees.

(1) Advantages of random forest algorithm

- Overfitting is less of an issue with random forest.
- It is fast but not in all cases.
- It is one of the most effective and versatile machine learning algorithms for wide variety of classification and regression tasks, as they are more robust to noise.
- It can be grown in parallel.
- It runs efficiently on large databases.

- It has higher classification accuracy.

(2) Applications of random forest algorithm

- It is used by banks to predict if a loan applicant is a likely high risk.
- It is used in the automobile industry to predict the failure or breakdown of a mechanical part.
- It is used in the healthcare industry to predict if a patient is likely to develop a chronic disease or not.
- It can also be used for regression tasks like predicting the average number of social media shares and performance scores.
- Recently it has also made its way into predicting patterns in speech recognition software and classifying images and texts.

9. Artificial neural networks algorithm

(1) Advantages of using artificial neural networks

- Easy to understand for professionals who do not want to dig deep into math-related complex machine learning algorithms.
- They are easy to conceptualize.
- They have the ability to identify all probable interactions between predictor variables.
- They have the ability to subtly identify complex nonlinear relationships that exists between independent and dependent variables.
- It is relatively easy to add prior knowledge to the model.

(2) Applications of artificial neural networks

Artificial neural networks are among the hottest machine learning algorithms in use today solving problems of classification to pattern recognition. They are extensively used in research and other application areas.

- Financial institutions use artificial neural networks algorithms to enhance the performance in evaluating loan applications, bond rating, target marketing, credit scoring. They are also used to identify instances of fraud in credit card transactions.
- BuzzFeed uses artificial neural network algorithms for image recognition to organize and search videos or photos.
- Google uses artificial neural networks for speech recognition, image recognition and other pattern recognition (handwriting recognition) applications. ANN's are used at Google to sniff out spam and for many other applications.
- Artificial neural networks find great applications in robotic factories for adjusting temperature settings, controlling machinery, diagnose malfunctions.

10. K-Nearest Neighbor (KNN) algorithm

KNN is one of the simplest machine learning algorithms based on supervised learning technique. KNN algorithm assumes the similarity between the new case/data and available cases and put the new case into the category that is most similar to the available categories. It stores all the available data and classifies a new data point based on the similarity. This means when new data appears then it can be easily classified into a well suite category by using KNN algorithm.

KNN is a non-parametric algorithm, which means it does not make any assumption on underlying data. It is also called a lazy learner algorithm because it does not learn from the training set immediately, instead it stores the dataset and at the time of classification, it performs an action on the dataset.

KNN algorithm can be used for regression as well as for classification but mostly it is used for the classification problems.

KNN algorithm at the training phase just stores the dataset and when it gets new data, then it classifies that data into a category that is much similar to the new data.

Example: Suppose we have an image of a creature that looks similar to cat and dog, but we want to know either it is a cat or dog. So for this identification, we can use the KNN algorithm, as it works on a similarity measure. Our KNN model will find the similar features of the new data set to the cats and dogs images and based on the most similar features it will put it in either cat or dog category.

New Words

confusing	[kən'fju:zɪŋ]	*adj.*难以理解的
binary	['baɪnəri]	*adj.*二态的；二元的
inspect	[ɪn'spekt]	*v.*检查；视察
epidemiology	[ˌepɪˌdi:mi'ɒlədʒi]	*n.*流行病学
methodology	[ˌmeθə'dɒlədʒi]	*n.*方法学；方法论
exemplify	[ɪg'zemplɪfaɪ]	*v.*是……的典范；举例说明
instinctual	[ɪn'stɪŋktʃuəl]	*adj.*本能的
constraint	[kən'streɪnt]	*n.*约束，限制
outlier	['aʊtlaɪə]	*n.*离群值；异常值
default	[di'fɔ:lt]	*v.*违约
random	['rændəm]	*adj.*随机的；任意的
ensemble	[ɒn'sɒmbl]	*n.*集成；全体，整体
poll	['pəʊl]	*v.*投票；轮询
versatile	['vɜ:sətaɪl]	*adj.*多用途的；多功能的
breakdown	['breɪkdaʊn]	*n.*故障
conceptualize	[kən'septʃuəlaɪz]	*vi.*概念化
subtly	['sʌtəli]	*adv.*巧妙地
non-parametric	[ˌnɒn pærə'metrɪk]	*adj.*无参数的

Phrases

be fit into	适合；融入
feature space	特征空间

multi-nominal logistic regression	多项逻辑回归
ordinal logistic regression	有序逻辑回归
linear relationship	线性关系
equal variance	等方差，同方差
normal distribution	正态分布
preventive measures	防护性措施，预防措施
credit scoring system	信用评分制度
risk management	风险管理
leaf node	叶节点
classification tree	分类树
regression tree	回归树
continuous variable	连续变量
binary variable	二元变量
data exploration	数据探索
missing value	缺失值
option pricing	期权定价
remote sensing	遥感
random forest	随机森林
random sample	随机样本
chronic disease	慢性病
bond rating	债券评级，债券分级
credit scoring	信用评分，资信评分
handwriting recognition	手写识别
diagnose malfunction	诊断故障
lazy learner	惰性学习器

Reference Translation

十大机器学习算法（II）

6. 逻辑回归算法

在逻辑回归算法用于分类任务而不是回归问题的意义上，此算法的名称有点令人困惑。这里的"回归"一词表示线性模型适合特征空间。该算法将逻辑函数应用于特征的线性组合，以根据预测变量来预测分类因变量的结果。它有助于根据给定的预测变量来估计属于分类因变量的特定级别的概率。

（1）逻辑回归的类型

- 二元逻辑回归：当分类响应有两种可能的结果（即是或否）时，逻辑回归最常用。示例：预测学生会不会通过考试，预测学生是否患有低血压或高血压，预测肿瘤是否癌变。
- 多项逻辑回归：分类响应有 3 个或更多个可能的结果，没有排序。示例：预测大多数美国公民使用哪种搜索引擎（如 Yahoo、Bing、Google 和 MSN）。
- 有序逻辑回归：分类响应有 3 个或更多个可能的自然排序结果。示例：客户如何根据 1～10 的等级对餐厅的服务和食物质量进行评分。

（2）逻辑回归的优点

- 易于检查且不那么复杂。
- 这是一种健壮的算法，因为自变量不必具有等方差或正态分布。
- 这些算法不假设因变量和自变量之间存在线性关系，因此也可以处理非线性效应。
- 它控制混淆并测试交互。

（3）逻辑回归的应用

- 它在流行病学领域中用于识别疾病的危险因素并相应地制定预防措施。
- 它用于将一组单词分类为名词、代词、动词、形容词。
- 它用于天气预报，如预测下雨的可能性。
- 它用于信用评分系统中的风险管理，以预测账户的违约情况。

7. 决策树算法

决策树是一种图形表示形式，它利用分支方法论来举例说明基于某些条件的决策的所有可能结果。在决策树中，内部节点代表对属性的测试，树的每个分支代表测试的结果，叶节点代表特定的类标签。即决定是在计算所有属性之后做出的。分类规则通过从根到叶节点的路径表示。

决策树有两种类型：分类树和回归树。根据目标变量的类型，它们也可以分为两类：连续变量决策树和二元变量决策树。目标变量有助于确定解决特定问题所需的决策树类型。

（1）决策树算法的优点

- 决策树是非常本能的，可以轻松地向任何人解释。
- 使用决策树算法时，数据类型不是一个约束，因为它们可以处理分类变量和数字变量。
- 决策树算法不需要对数据中的线性进行任何假设，因此可以在参数非线性相关的情况下使用。
- 这些算法在数据探索中很有用。
- 决策树对缺失值和异常值不敏感，因此有助于节省数据准备时间。

（2）决策树算法的应用

- 决策树是流行的机器学习算法之一，在金融中对期权定价非常有用。
- 遥感是基于决策树的模式识别的应用领域。
- 银行使用决策树算法，根据贷款申请人的违约概率对其进行分类。

8. 随机森林算法

随机森林算法是一种 go-to 算法，它使用袋装方法来创建一组带有数据随机子集的决策树。

在数据集的随机样本上对模型进行多次训练，以实现随机森林算法的良好的预测性能。在这种集成学习方法中，将随机森林中所有决策树的输出组合起来进行最终预测。随机森林算法的最终预测是通过轮询每个决策树的结果或仅通过在决策树中出现次数最多的预测来得出的。

（1）随机森林算法的优点

- 对于随机森林，过度拟合不是什么大问题。
- 速度很快，但并非在所有情况下都如此。
- 它是最有效和最通用的机器学习算法之一，适用于各种分类和回归任务，因为它们对噪声的抵抗力更强。
- 它可以并行生长。
- 它可以在大型数据库上高效运行。
- 分类精度更高。

（2）随机森林算法的应用

- 银行用来预测贷款申请人是否可能存在高风险。
- 在汽车工业中用于预测机械零件的失效或故障。
- 在医疗保健行业中用于预测患者是否可能患上慢性疾病。
- 它也可以用于回归任务，如预测社交媒体分享的平均数量和绩效得分。
- 最近，它也已用于预测语音识别软件中的模式以及对图像和文本进行分类。

9. 人工神经网络算法

（1）使用人工神经网络的优点

- 对于不想深入研究与数学相关的复杂机器学习算法的专业人员而言，它易于理解。
- 它们很容易概念化。
- 它们能够识别预测变量之间所有可能的相互作用。
- 它们能够巧妙地识别自变量和因变量之间存在的复杂非线性关系。
- 给模型添加先验知识相对容易。

（2）人工神经网络的应用

人工神经网络是当今使用的最热门的机器学习算法之一，可解决从分类到模式识别的问题。它们广泛用于研究和其他应用领域。

- 金融机构使用人工神经网络算法来增强评估贷款申请、债券评级、目标市场营销、信用评分的性能。它们还用于识别信用卡交易中的欺诈实例。
- BuzzFeed 使用人工神经网络算法进行图像识别，以组织和检索视频或照片。
- Google 把人工神经网络用于语音识别、图像识别和其他模式识别（手写识别）应用程序。Google 使用人工神经网络来发现垃圾邮件以及用于其他许多应用程序。
- 人工神经网络在机器人工厂中得到了广泛的应用，可用于调节温度设置、控制机械、诊断故障。

10. K 最近邻（KNN）算法

KNN 算法是基于监督学习技术的最简单的机器学习算法之一。KNN 算法假设新案例/数据与现有案例之间具有相似性，并将新案例放入与现有类别最相似的类别中。它存储所有现有数据，并根据相似度对新数据点进行分类。这意味着，当出现新数据时，可以使用 KNN 算法轻松地将其归入一个非常适当的类别。

KNN 是一种无参数算法，这意味着它不会对基础数据进行任何假设。它也称为惰性学习器算法，因为它不立即从训练集中学习，而是存储数据集，并且在分类时对数据集执行操作。

KNN 算法既可以用于回归也可以用于分类，但是大多数情况下用于分类问题。

训练阶段的 KNN 算法仅存储数据集，并在获取新数据时将其分类为与新数据非常相似的类别。

示例：假设我们有一个看起来类似于猫和狗的生物的图像，但是我们想知道它是猫还是狗。因此，对于这种识别，我们可以使用 KNN 算法，因为它适用于相似性度量。我们的 KNN 模型将找到新数据集与猫和狗图像相似的特征，并基于最相似的特征将其归入猫或狗类别。

Exercises

[Ex. 1] Answer the following questions according to Text A.

1. When does naive Bayes classifier algorithm come to the rescue?

2. What are the applications of naive Bayes classifier?

3. What is K-means?

4. What is K-means clustering algorithm used to do by most of the search engines like Yahoo, Google ?

5. What is Support Vector Machine (SVM)? How does it work?

6. How many categories of SVM are there? What are they?

7. What are the principles on which Apriori algorithm works?

8. What are the applications of Apriori algorithm?

9. What are the independent variables referred to? And why?

10. What are the applications of linear regression?

[Ex. 2] Fill in the following blanks according to Text B.

1. Logistic regression algorithm applies a logistic function to _____ to predict the outcome of a categorical dependent variable based on _____. It helps estimate the probability of falling into a specific level of _____ based on the given predictor variables.

2. The types of logistic regression listed in the passage are _____, _____, _____.

3. Logistic regression is applied in the field of epidemiology to identify _____ and plan accordingly for _____. It is used in _____ for risk management to predict the defaulting of an account.

4. A decision tree is _____ that makes use of branching methodology to exemplify all possible outcomes of a decision based on _____. In a decision tree, the internal node represents _____, each branch of the tree represents the outcome of the test and the leaf node represents _____.

5. There are two types of decision trees: _____ and _____. They can also be classified into two types based on the type of target variable: _____ and _____.

6. Decision tree algorithms do not require _____ on the linearity in the data and hence can be used in circumstances where _____ are non-linearly related. Decision tree algorithms are used by banks to _____ by their probability of _____.

7. Random forest is _____ that uses a bagging approach to create a bunch of decision trees with _____. The final prediction of the random forest algorithm is derived by _____ of each decision tree or just by _____that appears the most times in the decision trees.

8. Artificial neural networks _____ for professionals who do not want to dig deep into _____. They have the ability to subtly identify complex nonlinear relationships that exists between _____ and _____.

9. KNN is one of _____ machine learning algorithms based on _____. KNN algorithm assumes _____ between the new case/data and available cases and put the new case into the category that is most similar to _____.

10. KNN is _____, which means it does not make any assumption on _____. It is also called _____ because it does not learn from the training set immediately, instead it stores the dataset and at the time of classification, it _____ on the dataset.

[Ex. 3] Translate the following terms or phrases from English into Chinese.

1. converge
2. growth
3. rapid
4. regression
5. similarity
6. item set
7. feature vector
8. hierarchical clustering
9. iterative method
10. K-means clustering

1. _____
2. _____
3. _____
4. _____
5. _____
6. _____
7. _____
8. _____
9. _____
10. _____

[Ex. 4] Translate the following terms or phrases from Chinese into English.

1. *n.*相似度
2. *n.*约束，限制
3. *n.*离群值；异常值
4. *adj.*随机的；任意的
5. *n.*超集，扩展集，父集
6. 重要作用，重要角色
7. 搜索引擎

1. _____
2. _____
3. _____
4. _____
5. _____
6. _____
7. _____

8. 网页	8. _____
9. 分类树	9. _____
10. 手写识别	10. _____

[Ex. 5] Translate the following passage into Chinese.

Machine Learning Algorithms

Machine learning algorithms are pieces of code that help people explore, analyze and find meaning in complex data sets. Each algorithm is a finite set of unambiguous step-by-step instructions that a machine can follow to achieve a certain goal. In a machine learning model, the goal is to establish or discover patterns that people can use to make predictions or categorize information.

Machine learning algorithms use parameters that are based on training data—a subset of data that represents the larger set. As the training data expands to represent the world more realistically, the algorithm calculates more accurate results.

Different algorithms analyze data in different ways. They're often grouped by the machine learning techniques that they're used for: supervised learning, unsupervised learning and reinforcement learning. The most commonly used algorithms use regression and classification to predict target categories, find unusual data points, predict values and discover similarities.

1. Supervised learning

In supervised learning, algorithms make predictions based on a set of labeled examples that you provide. This technique is useful when you know what the outcome should look like.

For example, you provide a dataset that includes city populations by year for the past 100 years, and you want to know what the population of a specific city will be four years from now. The outcome uses labels that already exist in the data set: population, city and year.

2. Unsupervised learning

In unsupervised learning, the data points aren't labeled—the algorithm labels them for you by organizing the data or describing its structure. This technique is useful when you don't know what the outcome should look like.

For example, you provide customer data, and you want to create segments of customers who like similar products. The data that you're providing isn't labeled, and the labels in the outcome are generated based on the similarities that were discovered between data points.

3. Reinforcement learning

Reinforcement learning uses algorithms that learn from outcomes and decide which action to take next. After each action, the algorithm receives feedback that helps it determine whether the choice it made was correct, neutral or incorrect. It's a good technique to use for automated systems that have to make a lot of small decisions without human guidance.

For example, you're designing an autonomous car, and you want to ensure that it's obeying the law and keeping people safe. As the car gains experience and a history of reinforcement, it learns how to stay in its lane, go the speed limit and brake for pedestrians.

Unit 8

Text A

扫码听课文

Artificial Neural Networks

1. What are artificial neural networks

Artificial Neural Networks (ANN), also called Neural Networks (NN), are a set of algorithms, whose model is inspired by the human brain. Artificial neural networks are designed to recognise and distinguish between patterns. They interpret sensory data through a kind of machine perception, labelling or clustering raw input. The patterns they recognise are numerical, contained in vectors, into which data such as images, sound or text must be translated.

ANN contain a number of hidden layers through which the data is processed, allowing the machine to go "deep" in its learning, making connections and weighting input for the best results. The term "deep" refers to the number of layers in a neural network. While deep neural networks can have as many as two hundred layers, traditional neural networks only have a few, usually around three.

ANN were first developed in the 1950s in order to help computers behave as if they are interconnected brain cells, much like in a human brain. An artificial neural network is an attempt to simulate the network of neurons that make up the human brain. This is to give the computer the ability to learn and make its own decisions based on data.

The ability to learn is a fundamental aspect of intelligence. Although it can be a little difficult to precisely define what "learning" is in every context, a learning process in an artificial neural network can be explained as the problem being continuously updated in a network architecture so that the network can efficiently perform a specific task. Performance of a task is improved over time by iteratively updating the weights in the network. Artificial neural networks differ from traditional expert systems in the sense that they appear to learn underlying rules from the given collection of representative examples, this is what makes them exciting to work with.

2. How do artificial neural networks work

A typical neural network contains a large number of artificial neurons called units arranged in a series of layers. The input layer is where rules are predetermined and representative examples are given to show the ANN what the output should look like. Hidden layers are where the input is processed and "broken down". These layers are shown in the Figure 8-1.

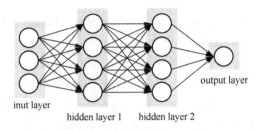

Figure 8-1   Neural network

The layers can be explained as follows.

- Input layer: It contains those units (artificial neurons) which receive input from the outside world on which network will learn, recognise about or otherwise process.
- Output layer: It contains units that respond to the information about how it's learned any task.
- Hidden layer: These units are in between input and output layers. The job of the hidden layer is to transform the input into something that output unit can use in some way.

Most neural networks are fully connected. That is to say each hidden neuron is fully linked to every neuron in its previous layer (input) and to the next layer (output) layer.See Figure 8-2.

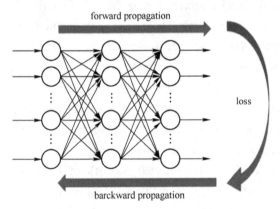

Figure 8-2   Information propagation in a neural network

Looking at an analogy may be helpful in understanding neural networks better. Learning in a neural network closely resembles to how we learn how to do things as humans—we perform an action and are either satisfied or dissatisfied with the result. Unsatisfactory results tend to cause a person to repeat a task until they succeed in achieving the desired result. Similarly, neural networks require a "supervisor" in order to describe what the desired result should be in response to the input.

Most deep learning methods use neural network architectures, which is why it is often referred to as deep neural networks. Deep learning models are trained by feeding them large amounts of labeled data and neural network architectures that learn features directly from the data without the need for manual input from a supervisor.

In deep learning networks, each layer of nodes learns on a distinct set of features based on the previous layer's output. The further you advance into the neural net, the more complex the features your nodes can recognise, since they aggregate and recombine features from the previous layer. This is known as feature hierarchy, which is a hierarchy of increasing complexity with each layer.

Based on the difference between the actual value and the predicted value, an error value which is also called cost function is calculated and sent back through the system.

Cost function: One half of the squared difference between actual and output value.

For each layer of the network, the cost function is analysed and used to adjust the threshold and weights for the next input. Our aim is to minimise the cost function. The lower the cost function, the closer the actual value to the predicted value. In this way, the error keeps becoming marginally lesser in each run as the network learns how to analyse values.

We feed the resulting data back through the entire neural network. The weighted synapses connecting input variables to the neuron are the only thing we have control over.

3. Types of neural networks

- Perceptron model neural network: A perceptron model neural network has two input units and one output unit with no hidden layers. These are also known as "single layer perceptrons".
- Radial basis function neural network: These networks are similar to the feed-forward neural network except radial basis function is used as the activation function of these neurons.
- Multilayer perceptron neural network: These networks use more than one hidden layer of neurons, unlike single layer perceptron. These are also known as deep feed-forward neural networks.
- Recurrent neural network: A type of neural network in which hidden layer neurons has self-connections. Recurrent neural networks possess memory. At any instance, hidden layer neuron receives activation from the lower layer as well as its previous activation value.
- Long Short Term Memory (LSTM) neural network: A type of neural network in which memory cell is incorporated into hidden layer neurons.
- Hopfield network: A fully interconnected network of neurons in which each neuron is connected to every other neuron. The network is trained with input pattern by setting a value of neurons to the desired pattern. Then its weights are computed. The weights are not changed. Once trained for one or more patterns, the network will converge to the learned patterns. It is different from other neural networks.
- Boltzmann machine neural network: These networks are similar to the Hopfield network, however some neurons are input, whereas others are hidden in nature. The weights are initialised randomly and learn through back propagation algorithm.
- Modular neural network: It is the combined structure of different types of the neural network like a multilayer perceptron, Hopfield network, recurrent neural network, which are incorporated into a single module of the network to perform independent subtasks of the whole neural network.
- Physical neural network: Physical neural networks comprise of electrically adjustable resistance material. It is used to emulate the function of synapse instead of software simulations performed in the neural network.

4. Different techniques of ANN

(1) Classification neural network

In a classification neural network, the network can be trained to classify any given patterns or datasets into a predefined class. It uses feed forward networks to do this.

(2) Prediction neural network

In a prediction neural network, the network can be trained to produce outputs that are expected from a given input. The network "learns" to produce outputs that are similar to the representation examples given in the input.

(3) Clustering neural network

The clustering neural network can be used to identify a unique feature of the data and classify them into different categories without any prior knowledge of the data.

The following networks are used for clustering.

● Competitive networks.

● Adaptive resonance theory networks.

● Kohonen self-organizing maps.

✎ New Words

inspire	[ɪn'spaɪə]	v.启发；激励
recognise	['rekəgnaɪz]	vt.辨识，认出，识别出
distinguish	[dɪ'stɪŋgwɪʃ]	v.区分，使有别于；辨别出
vector	['vektə]	n.向量，矢量
iteratively	['ɪtə,reitivli]	adv.迭代地
neuron	['njʊərɒn]	n.神经元；神经细胞
supervisor	['su:pəvaɪzə]	n.管理者，监督者
distinct	[dɪ'stɪŋkt]	adj.不同的；清楚的，明显的
threshold	['θreʃhəʊld]	n.阈值
synapse	['saɪnæps]	n.（神经元的）突触
perceptron	[pə'septrɒn]	n.感知器
feed-forward	[fi:d 'fɔ:wəd]	n.前馈
recurrent	[rɪ'kʌrənt]	adj.循环的，复现的；周期性的
possess	[pə'zes]	v.拥有；具备
activation	[,æktɪ'veɪʃn]	n.激活
propagation	[,prɒpə'geɪʃn]	n.传播，传输
modular	['mɒdjələ]	adj.模块化的
subtask	['sʌbtɑ:sk]	n.子任务
adjustable	[ə'dʒʌstəbl]	adj.可调整的，可调节的
resistance	[rɪ'zɪstəns]	n.电阻

| emulate | ['emjuleɪt] | *v.*模拟；仿效，模仿 |

✍ Phrases

underlying rule	潜在规则
artificial neuron	人工神经元
a series of	一系列；一连串
break down	分解
feature hierarchy	特征层次
actual value	实际值
predicted value	预测值
error value	误差值
cost function	成本函数
squared difference	平方差值
single layer perceptron	单层感知器
radial basis function	径向基函数
multilayer perceptron	多层感知器
deep feed-forward neural network	深度前馈神经网络
recurrent neural network	循环神经网络
memory cell	记忆细胞，记忆单元
adaptive resonance theory network	自适应共振理论网络
Kohonen self-organizing maps	科霍宁自组织映射

✍ Abbreviations

ANN (Artificial Neural Network)	人工神经网络
NN (Neural Network)	神经网络
LSTM (Long Short Term Memory)	长短期记忆

Reference Translation

人工神经网络

1. 什么是人工神经网络

人工神经网络（ANN）也称为神经网络（NN），是一组算法，其模型受人脑启发。人工神经网络旨在识别和区分模式。它们通过一种机器感知、标记或聚类原始输入来解释感官数据。它们识别的模式是数字化的，包含在向量中，必须将图像、声音或文本等数据转换为向量。

人工神经网络包含许多隐藏层，通过这些隐藏层处理数据，使机器能够"深入"学习、建立连接并加权输入以获得最佳结果。术语"深度"是指神经网络中的层数。虽然深度神经网络可以多达 200 层，但传统神经网络只有几层，通常在三层左右。

人工神经网络最初开发于 20 世纪 50 年代，目的是帮助计算机表现得好像它们是相互连接的脑细胞，就像在人脑中一样。人工神经网络试图模拟构成人脑的神经元网络。这是为了让计算机能够学习并根据数据做出自己的决定。

学习能力是智力的一个基本方面。虽然精确定义每种情况下的"学习"可能有点困难，但人工神经网络中的学习过程可以解释为网络架构中不断更新的问题，以便网络可以有效地执行特定任务。通过迭代更新网络中的权重，任务的性能会随着时间的推移而提高。人工神经网络与传统专家系统的不同之处在于，它们似乎是从给定的代表性示例集合中学习潜在规则，这便是使用人工神经网络的令人兴奋之处。

2. 人工神经网络是如何工作的

典型的神经网络包含大量称为单元的人工神经元，它们排列在一系列层中。输入层是预先确定规则的地方，并给出代表性的示例向人工神经网络展示输出层应该是什么样子。隐藏层是处理和"分解"输入的地方。这些层如图 8-1 所示。

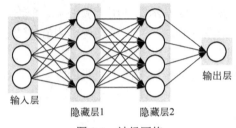

图 8-1　神经网络

这些层可以解释如下。

- 输入层：它包含那些从外部世界接收输入的单元（人工神经元），网络将在这些单元上学习、识别或进行其他处理。
- 输出层：它包含对关于如何学习任何任务的信息做出响应的单元。
- 隐藏层：这些单元位于输入层和输出层之间。隐藏层的工作是将输入转换为输出单元能够以某种方式使用的东西。

大多数神经网络是完全连接的。也就是说，每个隐藏神经元都与其前一层（输入）和下一层（输出）层中的每个神经元完全连接。如图 8-2 所示。

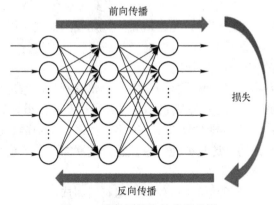

图 8-2　神经网络中的信息传播

进行类比可能有助于更好地理解神经网络。在神经网络中的学习与我们人类学习如何做事非常相似——我们执行一个动作，对结果要么满意要么不满意。不满意的结果往往会导致一个人重复一项任务，直到他们成功达到预期的结果。类似地，神经网络需要一个"管理者"来描述对输入的期望结果应该是什么。

大多数深度学习方法使用神经网络架构，这就是它通常被称为深度神经网络的原因。深度学习模型是通过向它们提供大量标记数据和神经网络架构来训练的，这些架构直接从数据中学习特征，而无须管理者手动输入。

在深度学习网络中，节点的每一层都根据前一层的输出来学习一组不同的特征。越深入神经网络，节点可以识别的特征就越复杂，因为它们聚合和重新组合来自前一层的特征。这被称为特征层次结构，它是层次增加复杂性也相应增加的结构。

根据实际值和预测值之间的差异，计算出一个误差值，也被称为成本函数，并通过系统返回。

成本函数：实际值与产出值的平方差的二分之一。

对于网络的每一层，都会分析成本函数并用于调整下一个输入的阈值和权重。我们的目标是最小化成本函数。成本函数越低，实际值越接近预测值。通过这种方式，随着网络学习如何分析值，每次运行中的误差就会变得越来越小。

我们通过整个神经网络将结果数据反馈回来。我们唯一可以控制的是输入变量连接到神经元的加权突触。

3. 神经网络的类型

- 感知器模型神经网络：感知器模型神经网络有两个输入单元和一个输出单元，没有隐藏层。也被称为"单层感知器"。
- 径向基函数神经网络：这类网络类似于前馈神经网络，只是径向基函数被用作这些神经元的激活函数。
- 多层感知器神经网络：与单层感知器不同，这类网络使用多个隐藏层的神经元。也称为深度前馈神经网络。
- 循环神经网络：一种隐藏层神经元具有自连接的神经网络。循环神经网络拥有记忆。在任何情况下，隐藏层神经元都会接收来自下层的激活以及其先前的激活值。
- 长短期记忆（LSTM）神经网络：一种将记忆细胞并入隐藏层神经元的神经网络。
- 霍普菲尔德（Hopfield）网络：一种完全互连的神经元网络，其中每个神经元都与其他每个神经元相连。通过将神经元的值设置为期望的模式，用输入模式训练网络。然后计算它的权重。权重不变。一旦针对一种或多种模式进行了训练，网络就会收敛到学习到的模式。它不同于其他神经网络。
- 玻尔兹曼机神经网络：这些网络类似于霍普菲尔德网络，但是一些神经元是输入的，而另一些实际上是隐藏的。权重随机初始化并通过反向传播算法学习。
- 模块化神经网络：它是不同类型的神经网络（如多层感知器、霍普菲尔德网络、循环神经网络）的组合结构，这些神经网络被整合到网络的单个模块中，以执行整个神经网络的独立子任务。
- 物理神经网络：物理神经网络由电可调电阻材料组成。它用于模拟突触的功能，而不是在神经网络中执行的软件模拟。

4. 人工神经网络的不同技术

（1）分类神经网络

在分类神经网络中，可以训练网络将任何给定的模式或数据集分类为预定义的类。它使用前馈网络来做到这一点。

（2）预测神经网络

在预测神经网络中，可以训练网络以产生从给定输入的预期输出。网络"学习"产生与输入中给出的表示示例相似的输出。

（3）聚类神经网络

聚类神经网络可用于识别数据的独特特征并将它们分为不同的类别，而无须对数据有任何先验知识。

以下网络用于聚类。

- 竞争网络。
- 自适应共振理论网络。
- 科霍宁自组织映射。

Text B

Machine Learning for Big Data

扫码听课文

In the past few years, more data has been produced than ever before. This data represents a gold mine in terms of commercial value and also important reference material for policy makers. But much of this value will stay untapped (or, worse, be misinterpreted) as long as the tools necessary for processing the staggering amount of information remain unavailable.

In this article, we'll look at how machine learning can give us insight into patterns in this sea of big data and extract key pieces of information hidden in it.

1. What is big data

Data consists of numbers, words, measurements and observations formatted in ways computers can process. Big data refers to vast sets of that data, either structured or unstructured.

The digital era presents a challenge for traditional data-processing software: information becomes available in such volume, velocity and variety that it ends up outpacing human-centered computation. And we can describe big data using these three "V"s: volume, velocity and variety. Volume refers to the scale of available data; velocity is the speed with which data is accumulated; variety refers to the different sources it comes from.

Two other Vs are often added to the aforementioned three: veracity and value. Veracity refers to the consistency and certainty (or lack thereof) in the sourced data, while value measures the usefulness of the data that's been extracted from the data received.

Good data analysis requires someone with business acumen, programming knowledge and a comprehensive skill set of math and analytic techniques. But how can a professional armed with traditional techniques sort through millions of credit card scores or billions of social media interactions? That's where machine learning comes in.

2. Big data meets machine learning

Machine learning algorithms become more effective as the size of training datasets grows. So when combining big data with machine learning, we benefit twice: the algorithms help us keep up with the continuous influx of data, while the volume and variety of the same data feeds the algorithms and helps them grow.

Let's look at how this integration process might work.

By feeding big data to a machine learning algorithm, we might expect to see defined and analyzed results, like hidden patterns and analytics, that can assist in predictive modeling.

For some companies, these algorithms might automate processes that were previously human-centered. But more often than not, a company will review the algorithm's findings and search them for valuable insights that might guide business operations.

While AI and data analytics run on computers that outperform humans by a vast margin, they lack certain decision-making abilities. Computers haven't got many characteristics inherent to humans, such as critical thinking, intention and the ability to use holistic approaches. Without an expert to provide the right data, the value of algorithm-generated results diminishes, and without an expert to interpret its output, suggestions made by an algorithm may compromise company decisions.

3. Machine learning applications for big data

Let's look at some real-life examples that demonstrate how big data and machine learning can work together.

(1) Cloud networks

A research firm has a large amount of medical data it wants to study, but in order to do so on-premises it needs servers, online storage, networking and security assets, all of which adds up to an unreasonable expense. Instead, the firm decides to invest in Amazon EMR, a cloud service that offers data-analysis models within a managed framework. Machine learning models of this sort include GPU-accelerated image recognition and text classification.

(2) Web scraping

Let's imagine that a manufacturer of kitchen appliances learns about market tendencies and customer satisfaction trends from a retailer's quarterly reports. In their desire to find out what the reports might have left out, the manufacturer decides to web scrape the enormous amount of existing data that pertains to online customer feedback and product reviews. By aggregating this data and feeding it to a deep learning model, the manufacturer learns how to improve and better describe its products, resulting in increased sales.

While web scraping generates a huge amount of data, it's worthwhile to note that choosing the sources for this data is the most important part of the process.

(3) Mixed initiative systems

The recommendation system that suggests titles on your Netflix homepage employs collaborative filtering: It uses big data to track your history (and everyone else's) and machine learning algorithms to decide what it should recommend next. This example demonstrates how big data and machine learning intersect in the arena of mixed-initiative systems, or human-computer interactions, whose results come from humans or machines taking initiative.

Similarly, smart car manufacturers implement big data and machine learning in the predictive analytics systems that run their products. Tesla cars, for example, communicate with their drivers and respond to external stimuli by using data to make algorithm-based decisions.

4. What to keep in mind

Achieving accurate results from machine learning has a few prerequisites. Apart from a well-built learning algorithm, you need clean data, scalable tools and a clear idea of what you want to achieve. While some might see these requirements as obstacles preventing their business from reaping the benefits of using big data with machine learning, in fact any business wishing to correctly implement this technology should invest in them.

(1) Data hygiene

Just as training for a sport can become dangerous for injury-prone athletes, learning from unsanitized or incorrect data can get expensive. Incorrectly trained algorithms produce results that will incur costs for a company and not save on them. Because mislabeled, missing or irrelevant data can impact the accuracy of your algorithm, you must be able to attest to the quality and completeness of your data sets as well as their sources.

(2) Practicing with real data

Suppose you want to create a machine learning algorithm but lack the massive amount of data required to train it. You might hear somewhere that derived computed data could be substituted for real data you generated. But beware: Because an ideal algorithm should solve a specific problem, it needs a specific type of data to learn from. Derived data rarely mimics the real data the algorithm needs to solve the problem, so using it almost guarantees that the trained algorithm will not fulfill its potential. Experimenting with real data offers the safest path.

(3) Knowing what you want to achieve

Don't let the hype around integrating machine learning with big data end up catapulting you into a poor understanding of the problem you want to solve. If you've pinpointed a complex problem but don't know how to use your data to solve it, you could wind up feeding inappropriate data to your algorithm or using correct data in inaccurate ways. To harness the power of big data, we recommend taking the time needed to create your own data before diving into an algorithm. That way you can educate yourself about your data, so when the time comes, you can use (and train) an algorithm appropriate to your problem.

(4) Scaling tools

Big data gives us access to more information, and machine learning increases our problem-solving capacity. Put together, the two present opportunities to scale entire businesses. To take advantage of this, we should also prepare our other tools (in the realms of finance, communication, etc.) for scaling.

5. Summary

In this article, we discussed the necessity of applying machine learning to big data analysis. By programming machines to interpret data too vast for humans to process alone, we can make decisions based on more accurate insights.

✎ New Words

untapped	[ˌʌn'tæpt]	adj.未开发的，未利用的
misinterpret	[ˌmɪsɪn't3:prət]	vt.误解，曲解
stagger	['stægə]	vt.使吃惊
unavailable	[ˌʌnə'veɪləbl]	adj.难以获得的；不能利用的
measurement	['meʒəmənt]	n.量度，测量
volume	['vɒlju:m]	n.大量
velocity	[və'lɒsəti]	n.高速
variety	[və'raɪəti]	n.多样，多样化
accumulate	[ə'kju:mjəleɪt]	v.积累，积聚
veracity	[və'ræsəti]	n.真实
value	['vælju:]	n.价值
comprehensive	[ˌkɒmprɪ'hensɪv]	adj.全面的；综合性的
continuous	[kən'tɪnjuəs]	adj.连续不断的
influx	['ɪnflʌks]	n.流入，注入；汇集
feed	[fi:d]	v.输送，供应
diminish	[dɪ'mɪnɪʃ]	v.（使）减小，减弱
tendency	['tendənsi]	n.倾向，趋势
title	['taɪtl]	n.标题 vt.加标题
track	[træk]	v.跟踪，追踪
prerequisite	[ˌpri:'rekwəzɪt]	n.先决条件，前提，必要条件
obstacle	['ɒbstəkl]	n.障碍（物）
incur	[ɪn'k3:]	vt.招致，引起
mislabel	[mɪs'leɪbl]	v.标记错误
irrelevant	[ɪ'reləvənt]	adj.不相干的；不恰当的
attest	[ə'test]	v.证实；证明
substitute	['sʌbstɪtju:t]	v.用……代替 n.代替者；替代物
fulfill	[fʊl'fɪl]	vt.履行，执行
hype	[haɪp]	n.炒作，天花乱坠的广告宣传 vt.大肆宣传；夸张地宣传

pinpoint	['pɪnpɔɪnt]	*vt.*确定
		*adj.*精确的，精准的；详尽的
realm	[relm]	*n.*领域，范围
usefulness	['ju:sflnəs]	*n.*有用，有益，有效

✎ Phrases

gold mine	金矿
policy maker	决策者，制定政策者
digital era	数字时代
business acumen	商业头脑，业务头脑
traditional technique	传统技术
hidden pattern	隐藏模式，隐含模式
predictive modeling	预测建模
work together	协同工作
online storage	在线存储
image recognition	图像识别
web scraping	网络抓取
product review	产品评论
mixed initiative system	混合主动系统
smart car manufacturer	智能汽车制造厂
keep in mind	牢记，谨记
derived computed data	派生的计算数据
end up	最终处于，到头来
wind up	（使自己）陷入，卷入，落得

Reference Translation

用于大数据的机器学习

在过去的几年里，所产生的数据比以往任何时候都多。这些数据就是一座具有商业价值的金矿，也是决策者的重要参考资料。但是，只要没有处理数量惊人的信息所需的工具，这些价值中的大部分将不会被利用（或者更糟的是，会被误解）。

在本文中，我们将研究机器学习如何让我们洞察大数据海洋中的模式并提取隐藏在其中的关键信息。

1. 什么是大数据

数据由数字、文字、测量值和观察值组成，其格式是计算机可以处理的。大数据是指大量结构化或非结构化的数据。

数字时代对传统数据处理软件提出了挑战：信息的数量、速度和多样性如此之大，最终超过以人为中心的计算。我们可以用这三个"V"来描述大数据：大量、高速和多样。大量是指可用数据的规模；高速是数据积累的速度；多样是指它有不同的来源。

经常会给前面提到的三个 V 添加另外两个 V：真实性和价值。真实性是指源数据的一致性和确定性（或缺乏一致性和确定性），而价值衡量从接收到的数据中提取的数据的有用性。

良好的数据分析需要具有业务头脑、编程知识以及全面的数学和分析技术技能。但是，拥有传统技术的专业人士如何对数百万张信用卡评分或对数十亿次社交媒体互动进行分类？这就是机器学习的用武之地。

2. 大数据遇上机器学习

随着训练数据集规模的增长，机器学习算法变得更加有效。因此，当将大数据与机器学习相结合时，我们会受益匪浅：算法帮助我们跟上不断涌入的数据，而相同数据的数量和种类则为算法提供支持并帮助它们成长。

让我们看看这个集成过程是如何工作的。

通过将大数据提供给机器学习算法，我们可能希望看到定义和分析的结果，如隐藏的模式和分析，这可以帮助预测建模。

对于一些公司来说，这些算法可能会使以前以人为中心的流程变得自动化。但通常情况下，公司会审查算法的发现，并从中寻找可能会指导业务运营的有价值的见解。

虽然人工智能和数据分析在计算机上运行的性能远远超过人类，但它们缺乏一定的决策能力。计算机尚未具备人类固有的许多特征，如批判性思维、意图和使用整体方法的能力。如果没有专家提供正确的数据，算法生成结果的价值就会降低，如果没有专家来解释其输出，算法提出的建议甚至可能会影响公司的决策。

3. 大数据的机器学习应用

让我们看一些现实生活中的例子，它们展示了大数据和机器学习如何协同工作。

（1）云网络

一家研究公司想要研究大量医疗数据，但为了在本地进行研究，它需要服务器、在线存储、网络和安全资产，所有这些加起来都增加了不理性的费用。于是，该公司决定投资Amazon EMR，这是一种在托管框架内提供数据分析模型的云服务。这类机器学习模型包括GPU 加速的图像识别和文本分类。

（2）网页抓取

假设厨房电器制造商从零售商的季度报告中了解市场趋势和客户满意度趋势。为了找出报告可能遗漏的内容，制造商决定通过网络抓取大量与在线客户反馈和产品评论有关的现有数据。通过汇总这些数据并将其提供给深度学习模型，制造商可以知道如何改进和更好地描述其产品，从而增加销售额。

虽然网络抓取会生成大量数据，但值得注意的是，选择这些数据的来源是该过程中最重要的部分。

（3）混合主动系统

在你的 Netflix 主页上推荐标题的推荐系统采用了协同过滤：它使用大数据来跟踪你的历史（和其他人的历史），并且用机器学习算法来决定接下来应该推荐什么。这个例子展示了大数据和机器学习如何在混合主动系统或人机交互领域相交，其结果来自人类或机器的主动性。

同样，智能汽车制造商在运行其产品的预测分析系统中实施大数据和机器学习。例如，特斯拉汽车通过使用数据做出基于算法的决策，与驾驶员进行交流并对外部刺激做出反应。

4. 注意事项

从机器学习中获得准确的结果有几个先决条件。除了构建良好的学习算法之外，你还需要干净的数据、可扩展的工具以及对想要实现的目标的清晰认识。虽然有些人可能会认为这些要求阻碍了他们的业务从使用机器学习的大数据中获益，但事实上，任何希望正确实施这项技术的企业都应该对它们进行投资。

（1）数据健康

正如一项运动的训练对容易受伤的运动员来说是危险的一样，从不干净的或不正确的数据中学习也会变得昂贵。训练不当的算法产生的结果会给公司带来支出，但不会节省成本。因为错误标记、丢失或不相关的数据会影响算法的准确性，所以必须能够证明数据集及其源的质量和完整性。

（2）实战演练

假设你想创建一个机器学习算法，但是缺少训练它所需的大量数据。你也许在某个地方听说过，派生的计算数据可以代替生成的真实数据。但是要注意：因为一个理想的算法应该解决一个特定的问题，所以它需要一个特定类型的数据来学习。派生的数据很少模仿算法解决问题所需的真实数据，因此使用它几乎可以保证训练后的算法不会发挥其潜力。用真实数据进行实验提供了最安全的途径。

（3）知道你想要达到的目标

不要让那些把机器学习和大数据结合起来的炒作最终使你对想解决的问题缺乏理解。如果你已经确定了一个复杂的问题，但不知道如何使用你的数据来解决它，那么你可能会向算法提供不适当的数据，或者以不准确的方式使用正确的数据。为了利用大数据的威力，我们建议在深入研究算法之前先花点时间创建自己的数据。这样，你就可以了解自己的数据，因此当时机成熟时，就可以使用（并训练）适合你的问题的算法。

（4）扩展工具

大数据让我们获得更多的信息，机器学习提高了我们解决问题的能力。两者结合在一起，就提供了扩展整个业务的机会。我们还应该准备其他工具（在金融、通信等领域）进行扩展。

5. 总结

在本文中，我们讨论了将机器学习应用于大数据分析的必要性。通过编程机器来解释人类无法单独处理的庞大数据，我们可以根据更准确的见解做出决策。

Exercises

[Ex. 1] Answer the following questions according to Text A.

1. What are artificial neural networks? What are they designed for?

2. What does ANN contain?

3. When and why were ANN first developed?

4. What are the layers a typical neural network contains?

5. How are deep learning models trained?

6. What are radial basis function neural networks?

7. What is Hopfield network? How is the network trained?

8. What do physical neural networks comprise of? What is it used to do?

9. What can a prediction neural network be trained to do?

10. What are the networks that are used for clustering?

[Ex. 2] Answer the following questions according to Text B.

1. What does data consist of? What does big data refer to?

2. What are the five "V"s used to describe big data?

3. What does veracity refer to? What does value do?

4. What might we expect by feeding big data to a machine learning algorithm?

5. If a research firm has a large amount of medical data it wants to study, what does it need to do so on-premises?

6. What does the manufacturer do by aggregating this data and feeding it to a deep learning model?

7. How do Tesla cars communicate with their drivers and respond to external stimuli?

8. Why must you be able to attest to the quality and completeness of your data sets as well as their sources?

9. Why does an ideal algorithm need a specific type of data to learn from?

10. What do we recommend to harness the power of big data?

[Ex. 3] Translate the following terms or phrases from English into Chinese.

1.	activation	1.	
2.	emulate	2.	
3.	inspire	3.	
4.	modular	4.	
5.	neuron	5.	
6.	artificial neuron	6.	
7.	deep feed-forward neural network	7.	
8.	Kohonen self-organizing maps	8.	
9.	recurrent neural network	9.	
10.	radial basis function	10.	

[Ex. 4] Translate the following terms or phrases from Chinese into English.

1.	*n.*感知器	1.	
2.	*n.*（神经元的）突触	2.	
3.	*n.*阈值	3.	
4.	*v.*积累，积聚	4.	
5.	*vt.*履行，执行	5.	

6. 实际值	6.	_____
7. 误差值	7.	_____
8. 输入层	8.	_____
9. 预测值	9.	_____
10. 平方差值	10.	_____

[Ex. 5] Translate the following passage into Chinese.

Deep Learning

Deep learning is a subset of machine learning concerned with large amounts of data with algorithms that have been inspired by the structure and function of the human brain. That is why deep learning models are often referred to as deep neural networks.

In deep learning, a computer model learns to perform classification tasks directly from images, text or sound. It performs a task repeatedly, and make a little tweak to improve the outcome. Deep learning models can exceed human-level performance. Models are trained by using a large set of labeled data and neural network architectures that contain many layers.

The most important part of a deep learning neural network is a layer of computational nodes called "neurons". Every neuron connects to all of the neurons in the underlying layer. Due to "deep learning" the neural network leverages at least two hidden layers. The addition of the hidden layers enables researchers to make more in-depth calculations.

How does the algorithm work then? The thing is, each connection has its weight or importance. But, with the help of the deep neural networks we can automatically find out the most important features for classification. This is performed with the help of the Activation Function that evaluates the way the signal should take for every neuron, just like in the case of a human brain.

Types of deep learning layers are as follows.

- The input layer of nodes receives the information and transfers it to the underlying nodes.
- Hidden node layers are the ones where the computations appear. This is the layer that uses those patterns of local contrast to fixate on things that resemble.
- In the output node layer, the results of the computations will show up. In this layer, the features will be applied to templates.

Unit 9

Text A

Pattern Recognition

扫码听课文

1. What is pattern recognition

Pattern recognition is the process of recognizing regularities in data by a machine that uses machine learning algorithms. In the heart of the process lies the classification of events based on statistical information, historical data or the machine's memory.

A pattern is a regularity in the world or in abstract notions. If we talk about books or movies, a description of a genre would be a pattern. If a person keeps watching black comedies, Netflix wouldn't recommend heartbreaking melodramas to him.

For the machine to search for patterns in data, it should be preprocessed and converted into a form that a computer can understand. Then, the researcher can use classification, regression, or clustering algorithms depending on the information available about the problem to get valuable results.

- Classification. In classification, the algorithm assigns labels to data based on the predefined features. This is an example of supervised learning.
- Clustering. An algorithm splits data into a number of clusters based on the similarity of features. This is an example of unsupervised learning.
- Regression. Regression algorithms try to find a relationship between variables and predict unknown dependent variables based on known data. It is based on supervised learning.

2. What should a pattern recognition system be able to do

If you want to assess how good or bad a pattern recognition system is, you need to pay attention to what it can do.

- Identify a familiar pattern quickly and accurately.
- Classify unfamiliar objects.
- Recognize shapes and objects from different angles.
- Uncover patterns and objects, even when partly hidden.
- Automatically recognize patterns.

3. How to build a pattern recognition system

To build a pattern recognition system, you need to choose a model and prepare the data. For pattern recognition, neural networks, classification algorithms (naive Bayes, decision tree, support vector machines) or clustering algorithms (K-means, mean shift, DBSCAN) are often used.

Next, you will work with data, divide it into three sets.

- Training set. We use the training set to train the model. You need to select representative samples and make the program process it using training rules. For example, if you are building a security system based on face recognition, you will need a variety of photos of your employees. All the relevant information will be extracted from these data. Generally, 80% of all the data makes a training data set.

- Validation set. This set is used to fine-tune the model. You use it to verify the effectiveness of the model. If the accuracy over the training data set increases, but the accuracy over the validation data set stays the same or decreases, then you're overfitting your model and you should stop training it.

- Testing set. Testing set is used to test whether the outputs given by the system are accurate. About 20% of the data is used for testing.

Note: do not confuse the validation set and testing set. The validation set is used to tune the parameters of the model while a testing set assesses its performance as a whole.

4. Components of a pattern recognition system

A pattern recognition system needs some input from the real world that it perceives with sensors. Such a system can work with any type of data,such as images, videos, numbers or texts.

Having received some information as the input, the algorithm performs preprocessing. That is segmenting something interesting from the background. For example, when you are given a group photo and a familiar face attracts your attention, this is preprocessing.

Preprocessing is tightly connected with enhancement. By this term, researchers understand an increase in the ability of a human or a system to recognize patterns even when they are vague. Imagine you are still looking at the same group photo but it is 20 years old. To make sure that the familiar face in the photo is really the person you know, you start comparing their hair, eyes and mouth. This is when enhancement steps into the game.

The next component is feature extraction. The algorithm uncovers some characteristic traits that are similar to more than one data sample.

The result of a pattern recognition system will be either a class assignment (if we use classification), cluster assignment (in case of clustering) or predicted values (if you apply regression). A pattern recognition system are shown in the Figure 9-1.

5. Types of pattern recognition models

There are three types of pattern recognition models.

(1) Statistical pattern recognition

This type of pattern recognition refers to statistical historical data when it learns from examples: it collects observations, processes them, and learns to generalize and apply these rules to new observations.

(2) Syntactic pattern recognition

It is also called structural pattern recognition because it relies on simpler subpatterns called primitives (for example, words). The pattern is described in terms of connections between the primitives, for example, words form sentences and texts.

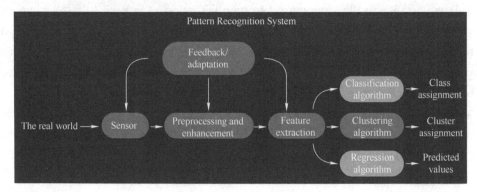

Figure 9-1　A Pattern Recognition System

(3) Neural pattern recognition

In neural pattern recognition, artificial neural networks are used. They can learn complex nonlinear input-output relations and adapt themselves to the data.

6. Pattern recognition process

Pattern recognition happens in two stages.

● First goes the explorative part. In general, the algorithm searches for patterns first.

● Next, there is the descriptive part, where the algorithm begins to categorize the found patterns.

The combination of the two is used to extract insights.

The process is as follows.

● First, you need to gather data.

● Then, you preprocess it and clean it from the noise.

● The algorithm examines the data and looks for relevant features or common elements.

● Then, these elements are classified or clustered.

● Each segment is analyzed for insights.

● Finally, the extracted insights are implemented in practice.

7. Where to use pattern recognition

Pattern recognition techniques are useful for solving classification problems, uncovering fraud, predicting volcanic eruptions, or diagnosing dangerous diseases with higher accuracy than humans.

(1) Image processing, segmentation and analysis

Pattern recognition is used for image processing. For example, a machine learning algorithm can recognize hundreds of bird species better than humans, even if the image is dark or noisy.

(2) Computer vision

One of the artificial neural networks that use pattern recognition for computer vision is Neural Talk that can generate descriptions of the environment in real-time.

(3) Speech recognition

Virtual assistants such as Alexa or Siri use speech recognition systems to process whole chunks of speech instead of working with separate words of phonemes.

(4) Fingerprint identification

A number of recognition methods have been used to perform fingerprint matching. The pattern recognition approach is widely used both for criminalistics and in your own smartphone. If you have a fingerprint lock on your phone, that is pattern recognition that steps into the game every time you unlock it.

(5) Stock market analysis

The stock market is hard to predict. However, even there, there are patterns that can be recognized and used. Modern apps for investors use AI to provide their users with consulting services.

(6) Medical diagnosis

Algorithms for pattern recognition trained on real data can be used for cancer diagnosis. These researchers have proposed an automatic breast cancer detection technique that gives a prediction accuracy of 99.86%. They use histopathology images from biopsy for feature extraction and apply an artificial neural network to produce the results.

New Words

pattern	['pætn]	n.模式
regularity	[ˌreɡju'lærəti]	n.规则性，规律性
classification	[ˌklæsɪfɪ'keɪʃn]	n.分类，类别
statistical	[stə'tɪstɪkl]	adj.统计的，统计学的
genre	['ʒɒnrə]	n.类型，种类
preprocess	[priː'prəʊses]	vt.预处理，预加工
valuable	['væljuəbl]	adj.有价值的
assign	[ə'saɪn]	v.指派；给予
relationship	[rɪ'leɪʃnʃɪp]	n.关系
accurately	['ækjərətli]	adv.准确地；精确地
unfamiliar	[ˌʌnfə'mɪliə]	adj.不熟悉的；不常见的
uncover	[ʌn'kʌvə]	v.揭示，发现
representative	[ˌreprɪ'zentətɪv]	adj.典型的；有代表性的
fine-tune	[ˌfaɪn 'tjuːn]	vt.微调
assess	[ə'ses]	v.评估；估价；估算
perceive	[pə'siːv]	v.察觉，发觉
sensor	['sensə]	n.传感器
background	['bækɡraʊnd]	n.背景
Component	[kəm'pəʊnənt]	n.部件，组件 adj.组成的，构成的

trait	[treɪt]	n.特点，特征，特性
generalize	['dʒenrəlaɪz]	v.概括，归纳
syntactic	[sɪn'tæktɪk]	adj.句法的
primitive	['prɪmətɪv]	n.原语
adapt	[ə'dæpt]	v.（使）适应/适合
explorative	[eks'plɒrətɪv]	adj.探索的
gather	['gæðə]	v.收集，搜集
real-time	[ˌriːl 'taɪm]	adj.实时的
phoneme	['fəʊniːm]	n.音位，音素
fingerprint	['fɪŋgəprɪnt]	n.指纹，指印
histopathology	[ˌhɪstəʊpə'θɒlədʒi]	n.组织病理学
biopsy	['baɪɒpsi]	n.活组织检查，活体检视

✎ Phrases

abstract notion	抽象概念
black comedy	黑色喜剧
convert into	转换为
split ... into	把……划分为……
training set	训练集
training rule	训练规则
face recognition	面部识别，人脸识别
validation set	验证集
testing set	测试集
feature extraction	特征提取
structural pattern recognition	结构化模式识别
neural pattern recognition	神经模式识别
common element	共同元素
fingerprint identification	指纹识别
fingerprint lock	指纹锁

✎ Abbreviations

DBSCAN (Density-Based Spatial Clustering of Applications with Noise) — 具有噪声的基于密度的聚类方法

Reference Translation

模式识别

1. 什么是模式识别

模式识别是使用机器学习算法的机器识别数据规律的过程。该过程的核心是基于统计信息、历史数据或机器记忆对事件进行分类。

模式是世界或抽象概念中的规律性。如果我们谈论书籍或电影，对一种类型的描述就是一种模式。如果一个人一直在看黑色喜剧，Netflix 不会向他推荐令人心碎的情节剧。

要让机器搜索数据中的模式，应该预处理数据并将其转换成计算机可以理解的形式。然后，研究人员可以根据有关问题的现有信息使用分类、回归或聚类算法来获得有价值的结果。

- 分类。在分类中，算法根据预定义的特征给数据分配标记。这是监督学习的一个例子。
- 聚类。根据特征的相似性，算法将数据分成多个集群。这是无监督学习的一个例子。
- 回归。回归算法试图找到变量之间的关系，并根据已知数据预测未知的因变量。它基于监督学习。

2. 模式识别系统能够做什么

如果你想评估一个模式识别系统的好坏，你需要注意它可以做什么。

- 快速准确地识别熟悉的模式。
- 对不熟悉的对象进行分类。
- 从不同角度识别形状和物体。
- 发现图案和物体，即使部分隐藏也可以。
- 自动识别模式。

3. 如何构建模式识别系统

要构建模式识别系统，你需要选择模型并准备数据。对于模式识别，经常使用神经网络、分类算法（朴素贝叶斯、决策树、支持向量机）或聚类算法（K-means、Mean Shift、DBSCAN）。

接下来，你将处理数据，把它分成三组。

- 训练集。我们使用训练集来训练模型。你需要选择具有代表性的样本并让程序使用训练规则对其进行处理。例如，如果你想构建一个基于人脸识别的安全系统，你将需要员工的各种照片。所有相关信息都将从这些数据中提取。通常，所有数据的 80% 构成了训练数据集。
- 验证集。该集合用于微调模型。可以使用它来验证模型的有效性。如果训练数据集的准确性增加，但验证数据集的准确性保持不变或下降，那么你的模型就过度拟合了，你应该停止训练它。
- 测试集。测试集用于测试系统给出的输出是否准确。大约 20% 的数据用于测试。

注意：不要混淆验证集和测试集。验证集用于调整模型的参数，而测试集则从整体上评估其性能。

4. 模式识别系统的组成部分

模式识别系统需要来自现实世界的一些输入，这些输入是通过传感器感知的。这样的系统可以处理任何类型的数据，如图像、视频、数字或文本。

接收到一些信息作为输入后，该算法执行预处理。这是从背景中分割出一些有趣的东西。例如，当给你一张合影，一张熟悉的面孔吸引了你的注意力时，这就是预处理。

预处理与增强密切相关。通过这个术语，研究人员理解人类或系统识别模式的能力有所提高，即使它们是模糊的。想象一下，你还在看同一张合影，但它已经是 20 年前的了。为了确保照片中熟悉的面孔确实是你认识的人，你开始比较他们的头发、眼睛和嘴巴。这就是增强操作。

下一个组件是特征提取。该算法揭示了一些与多个数据样本共有的特征。

模式识别系统的结果将是分类（如果使用分类）或分簇（在聚类的情况下）或预测值（如果应用回归）。模式识别系统如图 9-1 所示。

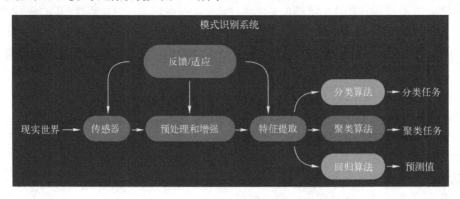

图 9-1 模式识别系统

5. 模式识别模型的类型

模式识别模型分为三种。

（1）统计模式识别

这种类型的模式识别指的是从示例中学习时的统计历史数据：它收集观察结果，处理它们，并学习概括这些规则并将其应用于新的观察结果。

（2）句法模式识别

它也称为结构化模式识别，因为它依赖于被称为原语（如单词）的更简单的子模式。模式是根据原语之间的联系来描述的，例如，单词构成句子和文本。

（3）神经模式识别

在神经模式识别中，使用人工神经网络。它们可以学习复杂的非线性输入-输出关系并适应数据。

6. 模式识别过程

模式识别分两个阶段进行。

● 首先是探索部分。该算法通常会先搜索模式。

● 接下来是描述部分，算法开始对找到的模式进行分类。

两者的结合用于获得洞察力。

该过程具体如下。

- 首先，你需要收集数据。
- 然后，你对其进行预处理并清除噪声。
- 算法检查数据并寻找相关特征或共同元素。
- 然后，对这些元素进行分类或聚类。
- 分析每一部分以获得洞察力。
- 最后，将提取的见解付诸实践。

7. 何处使用模式识别

模式识别技术可用于解决分类问题、发现欺诈、预测火山爆发或以比人类更高的准确度来诊断危险疾病。

（1）图像处理、分割和分析

模式识别用于图像处理。例如，机器学习算法可以比人类更好地识别数百种鸟类，即使图像很暗或杂乱。

（2）计算机视觉

将模式识别用于计算机视觉的人工神经网络之一是 Neural Talk，它可以实时生成环境描述。

（3）语音识别

Alexa 或 Siri 等虚拟助手使用语音识别系统来处理整个语音块，而不是处理每个单词的音素。

（4）指纹识别

许多识别方法已被用于执行指纹匹配。模式识别方法广泛用于犯罪调查和智能手机中。如果你的手机上有指纹锁，那就是每次解锁时都会使用模式识别。

（5）股市分析

股市很难预测。然而，即使在那里，也有可以识别和使用的模式。面向投资者的现代应用程序使用人工智能为其用户提供咨询服务。

（6）医学诊断

在真实数据上训练的模式识别算法可用于癌症诊断。这些研究人员已经提出了一种自动乳腺癌检测技术，其预测准确率为 99.86%。他们使用活检的组织病理学图像进行特征提取，并运用人工神经网络来产生结果。

Text B

Natural Language Processing

扫码听课文

1. What is natural language processing

Natural Language Processing (NLP) refers to the branch of computer science—and more specifically, the branch of artificial intelligence or AI—concerned with giving computers the ability to understand text and spoken words in much the same way human beings can.

NLP combines computational linguistics—rule-based modeling of human language—with statistical, machine learning and deep learning models. Together, these technologies enable computers to process human language in the form of text or voice data and to "understand" its full meaning.

NLP drives computer programs to translate text from one language to another, respond to spoken commands, and summarize large volumes of text rapidly—even in real time. There's a good chance you've interacted with NLP in the form of voice-operated GPS systems, digital assistants, speech-to-text dictation software, customer service chatbots and other consumer conveniences. NLP also plays a growing role in enterprise solutions that help streamline business operations, increase employee productivity and simplify mission-critical business processes.

2. NLP tasks

Human language is filled with ambiguities that make it incredibly difficult to write software that accurately determines the intended meaning of text or voice data. Homonyms, homophones, sarcasm, idioms, metaphors, grammar and usage exceptions, variations in sentence structure are just a few of the irregularities of human language that take humans years to learn. Programmers must teach natural language-driven applications to recognize and understand accurately from the start, if those applications are going to be useful.

Several NLP tasks break down human text and voice data in ways that help the computer make sense of what it's ingesting. Some of these tasks include are as follows.

- Speech recognition, also called speech-to-text, is the task of reliably converting voice data into text data. Speech recognition is required for any application that follows voice commands or answers spoken questions. What makes speech recognition especially challenging is the way people talk—quickly, slurring words together, with varying emphasis and intonation, in different accents, and often using incorrect grammar.

- Part of speech tagging, also called grammatical tagging, is the process of determining the part of speech of a particular word or piece of text based on its use and context. Part of speech identifies "make" as a verb in "I can make a paper plane," and as a noun in "What make of car do you own?"

- Word sense disambiguation is the selection of the meaning of a word with multiple meanings through a process of semantic analysis that determine the word that makes the most sense in the given context. For example, word sense disambiguation helps distinguish the meaning of the verb "make" in "make the grade" (achieve) vs. "make a bet" (place).

- Named Entity Recognition (NER), identifies words or phrases as useful entities. NER identifies "Kentucky" as a location or "Fred" as a man's name.

- Co-reference resolution is the task of identifying if and when two words refer to the same entity. The most common example is determining the person or object to which a certain pronoun refers (e.g., "she" = "Mary"), but it can also involve identifying a metaphor or an idiom in the text　(e.g., an instance in which "bear" isn't an animal but a large hairy person).

- Sentiment analysis attempts to extract subjective qualities from text, such as attitudes, emotions, sarcasm, confusion, suspicion.

- Natural language generation is sometimes described as the opposite of speech recognition or speech-to-text; it's the task of putting structured information into human language.

3. NLP tools and approaches

(1) Python and the Natural Language Toolkit (NLTK)

Python programing language provides a wide range of tools and libraries for accomplishing specific NLP tasks. Many of these are found in the Natural Language Toolkit or NLTK, an open source collection of libraries, programs and education resources for building NLP programs.

The NLTK includes libraries for many of the NLP tasks listed above, plus libraries for subtasks, such as sentence parsing, word segmentation, stemming and lemmatization (methods of trimming words down to their roots), and tokenization (for breaking phrases, sentences, paragraphs and passages into tokens that help the computer better understand the text). It also includes libraries for implementing capabilities such as semantic reasoning, the ability to reach logical conclusions based on facts extracted from text.

(2) Statistical NLP, machine learning and deep learning

The earliest NLP applications were hand-coded, rules-based systems that could perform certain NLP tasks, but couldn't easily scale to accommodate a seemingly endless stream of exceptions or the increasing volumes of text and voice data.

Statistical NLP combines computer algorithms with machine learning and deep learning models to automatically extract, classify, and label elements of text and voice data and then assign a statistical likelihood to each possible meaning of those elements. Today, deep learning models and learning techniques based on convolutional neural networks (CNN) and recurrent neural networks (RNN) enable NLP systems to "learn" as they work and extract ever more accurate meaning from huge volumes of raw, unstructured, and unlabeled text and voice data sets.

4. NLP use cases

Natural language processing is the driving force behind machine intelligence in many modern real-world applications. Here are a few examples.

- Spam detection: You may not think of spam detection as an NLP solution, but the best spam detection technologies use NLP's text classification capabilities to scan emails for language that often indicates spam or phishing. These indicators can include overuse of financial terms, characteristic bad grammar, threatening language, inappropriate urgency, misspelled company names, and more. Spam detection is one of a handful of NLP problems that experts consider "mostly solved".

- Machine translation: Google Translate is an example of widely available NLP technology at work. Truly useful machine translation involves more than replacing words in one language with words of another. Effective translation has to capture accurately the meaning and tone of the input language and translate it into text with the same meaning and desired impact in the output language. Machine translation tools are making good progress in terms of accuracy. A great way to test any machine translation tool is to translate text into one language and then back to the original. An oft-cited classic example: Not long ago, translating "The spirit is willing but the flesh is weak" from English to Russian and back yielded "The vodka is good but the meat is rotten." Today, the result is "The spirit desires, but the flesh is weak," which isn't perfect, but inspires much more confidence in the English-to-Russian translation.

● Virtual assistants and chatbots: Virtual assistants such as Apple Siri and Amazon Alexa use speech recognition to recognize patterns in voice commands and natural language generation to respond with appropriate action or helpful comments. Chatbots perform the same magic in response to typed text entries. The best of these also learn to recognize contextual clues about human requests and use them to provide even better responses or options over time. The next enhancement for these applications is question answering, the ability to respond to our questions.

● Social media sentiment analysis: NLP has become an essential business tool for uncovering hidden data insights from social media channels. Sentiment analysis can analyze language used in social media posts, responses, reviews, and more to extract attitudes and emotions in response to products, promotions and events.

● Text summarization: Text summarization uses NLP techniques to digest huge volumes of digital text and create summaries and synopses for indexes, research databases or busy readers who don't have time to read full text. The best text summarization applications use semantic reasoning and Natural Language Generation (NLG) to add useful context and conclusions to summaries.

✍ New Words

command	[kəˈmɑːnd]	n.&v.命令；控制
summarize	[ˈsʌməraɪz]	vt.总结，概述
dictation	[dɪkˈteɪʃn]	n.口述；听写
productivity	[ˌprɒdʌkˈtɪvəti]	n.生产率，生产力
ambiguity	[ˌæmbɪˈgjuːəti]	n.歧义；不明确，模棱两可
homonym	[ˈhɒmənɪm]	n.同形同音异义词
homophone	[ˈhɒməfəʊn]	n.同音异义词
sarcasm	[ˈsɑːkæzəm]	n.讥讽；嘲讽
idiom	[ˈɪdiəm]	n.习语，成语
metaphor	[ˈmetəfə]	n.隐喻，暗喻
irregularity	[ɪˌregjəˈlærəti]	n.不规则，无规律
reliably	[rɪˈlaɪəbli]	adv.可靠地，确实地
slur	[slɜː]	vt.含糊地说
emphasis	[ˈemfəsɪs]	n.重读
intonation	[ˌɪntəˈneɪʃn]	n.语调，声调
disambiguation	[ˌdɪsæmˌbɪgjʊˈeɪʃən]	n.消歧，消除模棱两可情况
opposite	[ˈɒpəzɪt]	n.对立物
		adj.对面的；截然相反的
stem	[stem]	n.词干
lemmatization	[lemətaɪˈzeɪʃn]	n.词形还原
endless	[ˈendləs]	adj.无穷尽的；无止境的

phishing	[ˈfɪʃɪŋ]	n.网络钓鱼，网络仿冒
overuse	[ˌəʊvəˈjuːs]	vt.过度使用
inappropriate	[ˌɪnəˈprəʊpriət]	adj.不恰当的，不适宜的
tone	[təʊn]	n.语调；风格
progress	[ˈprəʊgres]	n.进步；前进；进展
contextual	[kənˈtekstʃuəl]	adj.上下文的；与语境有关的
clue	[kluː]	n.线索；提示
post	[pəʊst]	n.（论坛等的）帖子
digest	[daɪˈdʒest]	v.消化；理解
	[ˈdaɪdʒest]	n.摘要；文摘
synopsis	[sɪˈnɒpsɪs]	n.摘要，梗概；大纲

Phrases

concern with	与……有关；关心……
rule-based modeling	基于规则建模
mission-critical business process	执行关键任务的业务流程
be filled with	充满着
grammatical tagging	语法标注
co-reference resolution	指代消歧，共指解析
structured information	结构化信息
put ... into	把……译成……
sentence parsing	句子解析
word segmentation	词切分
semantic reasoning	语义推理
spam detection	垃圾邮件检测
machine translation	机器翻译

Abbreviations

NER (Named Entity Recognition)	命名实体识别
NLTK (Natural Language Toolkit)	自然语言工具包
NLG (Natural Language Generation)	自然语言生成

Reference Translation

自然语言处理

1. 什么是自然语言处理

自然语言处理（NLP）是计算机科学的一个分支，更具体地说，是人工智能（AI）的一个分支，它致力于让计算机能够像人类一样理解文本和话语。

NLP 将计算语言学（基于规则的人类语言建模）与统计、机器学习和深度学习模型相结合。这些技术使计算机能够以文本或语音数据的形式处理人类语言，并"理解"其全部含义。

NLP 促使计算机程序将文本从一种语言翻译成另一种语言、响应口头命令并快速汇总大量文本——甚至是实时的。你很有可能已经通过语音操作的 GPS 系统、数字助理、语音到文本听写软件、客户服务聊天机器人和其他消费者便利设施与 NLP 进行了互动。NLP 在简化业务运营、提高员工生产力和简化关键任务业务流程的企业解决方案中也发挥着越来越大的作用。

2. NLP 的任务

人类语言充满了歧义，这使得编写能够准确确定文本或语音数据含义的软件变得异常困难。同形同音异义词、同音异义词、讽刺、成语、隐喻、语法和用法异常、句子结构的变化只是人类语言不规则中的一小部分，人需要数年时间才能学会。如果要让应用程序可用，程序员必须从一开始就教会自然语言驱动的应用程序准确地识别和理解。

一些 NLP 的任务是以有助于计算机理解它所摄取的内容的方式来分解人类文本和语音数据。其中一些任务如下。

- 语音识别，也称为语音转文本，任务是将语音数据可靠地转换为文本数据。任何遵循语音命令或回答口头问题的应用程序都需要语音识别。使语音识别特别具有挑战性的是人们说话的方式，如语速快、吐字含糊不清、重音和语调不同、口音不同，并且经常使用不正确的语法。

- 词性标注，也称为语法标注，是根据特定单词或文本的用途和上下文确定其词性的过程。"make"在"I can make a paper plane"（我能做一架纸飞机）中是动词，在"What make of car do you own?"（你有什么牌子的汽车？）中是名词。

- 词义消歧是通过语义分析过程选择具有多重含义的词的含义，该过程可以确定在给定上下文中最有意义的词。例如，词义消歧有助于区分"make the grade"（达标，成功）与"make a bet"（打赌，下赌注）中动词"make"的含义。

- 命名实体识别（NER）将单词或短语识别为有用的实体。NER 将"Kentucky"标识为一个地点，将"Fred"标识为一个人的名字。

- 指代消歧的任务是识别两个词是否以及何时指同一实体。最常见的例子是确定某个代词所指的人或物（例如，"她"="玛丽"），但它也可能涉及识别文本中的隐喻或成语（例如，"熊"不是动物，而是一个毛茸茸的大个子）。

- 情感分析试图从文本中提取主观品质，如态度、情绪、讽刺、困惑、怀疑。

- 自然语言生成有时被描述为语音识别或语音到文本的对立面。它的任务是将结构化信息转化为人类语言。

3. NLP 工具和方法

（1）Python 和自然语言工具包（NLTK）

Python 编程语言提供了广泛的工具和库来完成特定的 NLP 任务。其中许多都可以在自然语言工具包（NLTK）中找到，NLTK 是用于构建 NLP 程序的库、程序和教育资源的开源集。

NLTK 包括用于上面列出的许多 NLP 任务的库以及用于子任务的库，如句子解析、词切分、词干提取、词形还原（将单词修剪到其词根的方法）和标记化（把词组、句子、段落和篇章分解为有助于计算机更好地理解文本的标记）。它还包括用于实现语义推理等功能的库，语义推理能够根据从文本中提取的事实得出逻辑结论。

（2）统计 NLP、机器学习和深度学习

最早的 NLP 应用程序是手工编码的、基于规则的系统，可以执行某些 NLP 任务，但无法轻松扩展以适应似乎无穷无尽的异常流或不断增加的文本和语音数据量。

统计 NLP 将计算机算法与机器学习和深度学习模型相结合，自动提取、分类和标记文本与语音数据的元素，然后为这些元素的每个可能含义分配统计可能性。现在，基于卷积神经网络（CNN）和循环神经网络（RNN）的深度学习模型与学习技术使 NLP 系统能够一边工作一边"学习"，并从大量原始、非结构化以及未标记的文本和语音数据集中提取更准确的含义。

4. NLP 用例

自然语言处理是许多现代现实世界应用中机器智能背后的驱动力。一些示例如下。

- 垃圾邮件检测：你可能不认为垃圾邮件检测是 NLP 解决方案，但最好的垃圾邮件检测技术正是使用 NLP 的文本分类功能来扫描电子邮件，以查找经常标明是垃圾邮件或网络钓鱼的语言。这些标志可能包括财务术语的过度使用、典型的语法错误、威胁性的语言、不恰当的紧迫情况、拼错的公司名称等。专家认为垃圾邮件检测是 NLP "大部分已解决"的少数问题之一。

- 机器翻译：谷歌翻译是广泛使用的 NLP 技术的一个例子。真正有用的机器翻译不仅仅是用一种语言的单词替换另一种语言的单词。有效的翻译必须准确捕捉输入语言的含义和语气，并将其翻译成与输出语言具有相同含义和预期影响的文本。机器翻译工具在准确性方面取得了良好的进展。测试任何机器翻译工具的一个好方法是将文本翻译成一种语言，然后再翻译回原始语言。一个经常被引用的经典例子：不久前，将英语 "The spirit is willing but the flesh is weak.(心有余而力不足。)" 翻译成俄语，然后再从俄语翻译成英语，得到的句子是 "The vodka is good but the meat is rotten.(伏特加不错，但肉烂了。)"。如今机器翻译的结果是 "The spirit desires, but the flesh is weak.(精神渴望，但肉体软弱。)"，这并不完美，但更激发了人们对英俄翻译的信心。

- 虚拟助手和聊天机器人：Apple Siri 和 Amazon Alexa 等虚拟助手使用语音识别来识别语音命令与自然语言生成中的模式，以适当的操作或有用的评论进行响应。聊天机器人执行相同的魔法功能来响应输入的文本条目。其中最好的机器人还学习识别有关人类请求的上下文线索，并随着时间的推移使用它们提供更好的响应或选项。这些应用程序的下一个增强功能是问答功能，即能够回答我们的问题。

- 社交媒体情绪分析：自然语言处理已成为从社交媒体渠道发现隐藏数据见解的重要业务工具。情感分析可以分析社交媒体帖子、回复、评论等中使用的语言，以提取对产品、促销和事件的态度与情感。

- 文本摘要：文本摘要使用自然语言的处理技术消化大量数字文本，并为索引、研究数据库或没有时间阅读全文的忙碌的读者创建摘要和概要。最好的文本摘要应用程序使用语义推理和自然语言生成（NLG）为摘要添加有用的上下文与结论。

Exercises

[Ex. 1] Answer the following questions according to Text A.

1. What is pattern recognition?

2. What do regression algorithms try to do?

3. What are the three sets data is divided into? What are they used to do respectively?

4. What are the components of a pattern recognition system?

5. What will the result of a pattern recognition system be?

6. How many types of pattern recognition models are there? What are they?

7. How does pattern recognition happen?

8. What are pattern recognition techniques useful for?

9. What is one of the artificial neural networks that use pattern recognition for computer vision?

10. What have these researchers proposed? What do they do?

[Ex. 2] Fill in the following blanks according to Text B.

1. Natural Language Processing (NLP) refers to the branch of _____—and more specifically, the branch of _____—concerned with giving computers the ability to understand _____ and _____ in much the same way human beings can.

2. Human language is filled with ambiguities that make it _____ to write software that accurately determines _____ of text or voice data. Programmers must teach natural language-driven applications to _____ and _____ accurately from the start, if those applications are going to be useful.

3. Speech recognition, also called _____, is the task of reliably converting _____ into text data. What makes speech recognition especially challenging is the way people talk—quickly, _____, _____ and intonation, _____, and often using incorrect grammar.

4. Word sense disambiguation is the selection of the meaning of a word with _____ through a process of _____ that determine the word that makes the most sense _____.

5. Sentiment analysis attempts to extract subjective qualities from text, such as attitudes, _____, sarcasm, _____, suspicion Natural language generation is sometimes described as the opposite of _____ or _____; it's the task of putting _____ into human language.

6. Statistical NLP combines computer algorithms with _____ and _____ to automatically extract, _____, and label elements of text and voice data and then assign _____ to each possible meaning of those elements.

7. The best spam detection technologies use _____ to scan emails for language that often indicates spam or _____. These indicators can include _____, _____, threatening language, inappropriate urgency, _____, and more.

8. Effective translation has to capture accurately _____ of the input language and translate it into text _____ with the same meaning and desired impact _____. A great way to test any machine translation tool is to translate text into one language and then _____.

9. Virtual assistants such as Apple Siri and Amazon Alexa use _____ to recognize patterns in _____ and natural language generation to respond _____ or

helpful comments. Chatbots perform the same magic in response to _____.

10. Text summarization uses NLP techniques to digest _____and create _____ for indexes, _____ or busy readers who don't have time to read full text.

[Ex. 3] Translate the following terms or phrases from English into Chinese.

1. fingerprint   1. _____

2. generalize   2. _____

3. pattern   3. _____

4. preprocess   4. _____

5. regularity   5. _____

6. neural pattern recognition   6. _____

7. structural pattern recognition   7. _____

8. training rule   8. _____

9. validation set   9. _____

10. feature extraction   10. _____

[Ex. 4] Translate the following terms or phrases from Chinese into English.

1. *v.*（使）适应/适合   1. _____

2. *v.*指派；给予   2. _____

3. *n.*关系   3. _____

4. *n.*不规则，无规律   4. _____

5. *n.*生产率，生产力   5. _____

6. 测试集   6. _____

7. 训练集   7. _____

8. 机器翻译   8. _____

9. 垃圾邮件检测   9. _____

10. 指纹识别   10. _____

[Ex. 5] Translate the following passage into Chinese.

Image Pattern Recognition

Image recognition is a variation of OCR aimed at understanding what is on the picture. In contrast with OCR, image recognition to recognize what is depicted on the input images during image processing. Basically, instead of "recognizing" is "describes" the picture so that it would be searchable and comparable with the other images.

The main algorithms at work in image recognition are a combination of unsupervised and supervised machine learning algorithms.

The first supervised algorithm is used to train the model on the labeled datasets, i.e., examples of the depiction of the objects. Then the unsupervised algorithm is used to explore an input image.

After this, a supervised algorithm kicks in and classifies the patterns as related to the particular category of objects (for example, an ink pen).

There are two main use cases for image recognition.

- Visual search features are widely used in search engines and e-commerce marketplaces. It works the same way as an alphanumeric search query only with images. The other part is image metadata and also additional textual input. This information is used to increase the efficiency of the results and to filter the selection of options according to the context. For example, such technologies are widely applied by Google Search and Amazon.

- Face detection is widely used in social network services, such as Facebook and Instagram. The same technology is used by law enforcement to find a person of interest or criminals on the run. The technical process behind face detection is more intricate than simple object recognition. To recognize the appearance of a certain person, the algorithm needs to have a specialized labeled sample set. However, due to privacy limitations, these features are usually optional and require user consent.

Unit 10

Text A

Intelligent Agents in Artificial Intelligence

扫码听课文

1. What is an Intelligent Agent (IA)

An intelligent agent is an entity that makes a decision, and that enables artificial intelligence to be put into action. It can also be described as a software entity that conducts operations in the place of users or programs after sensing the environment. It uses actuators to initiate action in that environment.

This agent has some level of autonomy that allows it to perform specific, predictable and repetitive tasks for users or applications. It's also termed as "intelligent" because of its ability to learn during the process of performing tasks.

The two main functions of intelligent agents include perception and action. Perception is done through sensors while actions are initiated through actuators.

Intelligent agents consist of sub-agents that form a hierarchical structure. Lower-level tasks are performed by these sub-agents. The higher-level agents and lower-level agents form a complete system that can solve difficult problems through intelligent behaviors or responses.

2. Characteristics of intelligent agents

Intelligent agents have the following characteristics.

- They have some level of autonomy that allows them to perform certain tasks on their own.
- They have a learning ability that enables them to learn even as tasks are carried out.
- They can interact with other entities such as agents, humans and systems.
- They can accommodate new rules incrementally.
- They exhibit goal-oriented habits.
- They are knowledge-based. They use knowledge regarding communications, processes and entities.

3. The structure of intelligent agents

The structure of IA consists of three main parts: architecture, agent function and agent program.

Architecture: This refers to machinery or devices that consists of actuators and sensors. The intelligent agent executes on this machinery. Examples include a personal computer, a car or a camera.

Agent function: The agent function is a mathematical function that maps a sequence of perceptions into action.

Agent program: The agent program implements or executes the agent function. The agent function is produced through the agent program's execution on the physical architecture.

4. Categories of intelligent agents

There are 5 main categories of intelligent agents. The grouping of these agents is based on their capabilities and level of perceived intelligence.

(1) Simple reflex agents

These agents perform actions using the current percept, rather than the percept history. The condition-action rule is used as the basis for the agent function. In this category, a fully observable environment is ideal for the success of the agent function.

(2) Model-based reflex agents

Unlike simple reflex agents, model-based reflex agents consider the percept history in their actions. The agent function can still work well even in an environment that is not fully observable. These agents use an internal model that determines the percept history and effect of actions. They reflect on certain aspects of the present state that have been unobserved.

(3) Goal-based agents

These agents have higher capabilities than model-based reflex agents. Goal-based agents use goal information to describe desirable capabilities. This allows them to choose among various possibilities. These agents select the best action that enhances the attainment of the goal.

(4) Utility-based agents

These agents make choices based on utility. They are more advanced than goal-based agents because of an extra component of utility measurement. Using a utility function, a state is mapped against a certain measure of utility. A rational agent selects the action that optimizes the expected utility of the outcome.

(5) Learning agents

These are agents that have the capability of learning from their previous experience. Learning agents have the following elements.

The learning element: This element enables learning agents to learn from previous experiences.

The critic: It provides feedback on how the agent is doing.

The performance element: This element decides on the external action that needs to be taken.

The problem generator: This acts as a feedback agent that performs certain tasks such as making suggestions (new) and keeping history.

5. How do intelligent agents work

Intelligent agents work through three main components: sensors, actuators and effectors. Getting an overview of these components can improve our understanding of how intelligent agents work.

Sensors: These are devices that detect any changes in the environment. This information is sent to other devices. In artificial intelligence, the environment of the system is observed by intelligent agents through sensors.

Actuators: These are components through which energy is converted into motion. They perform the role of controlling and moving a system. Examples include rails, motors and gears.

Effectors: The environment is affected by effectors. Examples include legs, fingers, wheels, display screen and arms.

The Figure 10-1 shows how these components are positioned in the AI system.

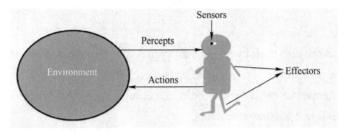

Figure 10-1　The components are positioned in the AI system

Inputs (percepts) from the environment are received by the intelligent agent through sensors. This agent uses artificial intelligence to make decisions using the acquired information/ observations. Actions are then triggered through actuators. As shown in Figure 10-2, future decisions will be influenced by percept history and past actions.

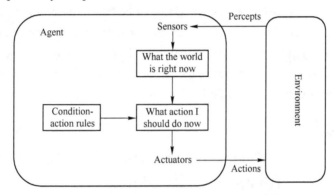

Figure 10-2　The working process of intelligent agents

6. Applications of intelligent agents

Intelligent agents in artificial intelligence have been applied in many real-life situations.

(1) Information search, retrieval and navigation

Intelligent agents enhance the access and navigation of information. This is achieved through the search of information using search engines. The internet consists of many data objects that may take users a lot of time to search for a specific data object. Intelligent agents perform this task on behalf of users within a short time.

(2) Repetitive office activities

Some companies have automated certain administrative tasks to reduce operating costs. Some of the functional areas that have been automated include customer support and sales. Intelligent agents have also been used to enhance office productivity.

(3) Medical diagnosis

Intelligent agents have also been applied in healthcare services to improve the health of patients.

In this case, the patient is considered as the environment. The computer keyboard is used as the sensor that receives data on the symptoms of the patient. The intelligent agent uses this information to decide the best course of action. Medical care is given through actuators such as tests and treatments.

(4) Environment cleaning

Intelligent agents are also used to enhance efficiency and cleanness in environment cleaning. In this case, the environment can be a room table or carpet. Some of the sensors employed in vacuum cleaning include cameras, bump sensors and dirt detection sensors. Action is initiated by actuators such as brushes, wheels and vacuum extractors.

(5) Autonomous driving

Intelligent agents enhance the operation of self-driving cars. In autonomous driving, various sensors are employed to collect information from the environment. These include cameras, GPS and radar. In this application, the environment can be pedestrians, other vehicles, roads or road signs. Various actuators are used to initiate actions. For example, brakes are used to bring the car to a stop.

7. Conclusion

Intelligent agents make work easier by performing certain time-consuming and difficult tasks on behalf of systems or users. These agents are making the automation of certain tasks possible.

With the increasing progress of technology, the development of intelligent agents will be strengthened. They will be further translated into complex AI driven devices to address current global challenges. There seems to be no limit to this interesting technology.

New Words

conduct	[kən'dʌkt]	v.引导，组织，实施
actuator	['æktʃʊeɪtə]	n.执行器
initiate	[ɪ'nɪʃieɪt]	vt.启动，开始，发起
autonomy	[ɔ:'tɒnəmi]	n.自主（权）
predictable	[prɪ'dɪktəbl]	adj.可预测的，可预报的
repetitive	[rɪ'petətɪv]	adj.重复的
sub-agent	[sʌb 'eɪdʒənt]	n.子智能体
response	[rɪ'spɒns]	n.响应，反应
accommodate	[ə'kɒmədeɪt]	v.适应
goal-oriented	[gəul 'ɔ:riəntid]	adj.面向目标的，目标导向的
habit	['hæbɪt]	n.习惯
device	[dɪ'vaɪs]	n.设备，装置
execute	['eksɪkju:t]	v.执行，实施
reflex	['ri:fleks]	n.反射作用；反应能力

observable	[əbˈzɜːvəbl]	adj.可观察的
utility	[juːˈtɪləti]	n.效用，功用
extra	[ˈekstrə]	adj.额外的
optimize	[ˈɒptɪmaɪz]	vt.使最优化
feedback	[ˈfiːdbæk]	n.反馈
effector	[ɪˈfektə]	n.效应器
situation	[ˌsɪtʃuˈeɪʃn]	n.情况，情景
retrieval	[rɪˈtriːvl]	n.检索
keyboard	[ˈkiːbɔːd]	n.键盘
treatment	[ˈtriːtmənt]	n.治疗
autonomous	[ɔːˈtɒnəməs]	adj.自治的，自主的
radar	[ˈreɪdɑː]	n.雷达
brake	[breɪk]	n.制动器
time-consuming	[ˈtaɪm kənˌsjuːmɪŋ]	adj.费时的，耗时的

✍ Phrases

be described as	被描述为
higher-level agent	高级智能体
lower-level agent	低级智能体
model-based reflex agent	基于模型的反射型智能体
goal-based agent	基于目标的智能体
utility-based agent	基于效用的智能体
data object	数据对象
operating cost	运营成本
customer support	客户支持
bump sensor	碰撞传感器
dirt detection sensor	污垢检测传感器
vacuum extractor	真空吸尘器
autonomous driving	自动驾驶
road sign	交通标志，路标
AI-driven device	由人工智能驱动的设备

✍ Abbreviations

| IA (Intelligent Agent) | 智能体 |

Reference Translation

人工智能中的智能体

1. 什么是智能体 (IA)

智能体是做出决策并使人工智能付诸行动的实体。也可以描述为感知环境后引导用户或程序进行操作的软件实体。它使用执行器在该环境中启动动作。

智能体具有一定程度的自主权，允许它为用户或应用程序执行特定的、可预测的和重复的任务。它也被称为"智能"，因为它具有在执行任务过程中学习的能力。

智能体的两个主要功能包括感知和行动。感知是通过传感器完成的，而行动是通过执行器发起的。

智能体由形成层次结构的子体组成。较低级别的任务由这些子体执行。高级智能体和低级智能体形成一个完整的系统，可以通过智能行为或响应解决难题。

2. 智能体的特点

智能体具有以下特征。

- 它们有一定程度的自主权，允许它们自己执行某些任务。
- 它们具有学习能力，使它们即使在执行任务时也能学习。
- 它们可以与其他实体交互，如智能体、人和系统。
- 它们可以逐渐适应新规则。
- 它们表现出以目标为导向的习惯。
- 它们是基于知识的。它们使用有关通信、流程和实体的知识。

3. 智能体的结构

智能体由架构、智能体函数和智能体程序三个主要部分组成。

架构：这是指由执行器和传感器组成的机器或设备。智能体在这台机器上执行。示例包括个人计算机、汽车或相机。

智能体函数：智能体函数是一个数学函数，它将一系列感知映射为行动。

智能体程序：智能体程序实现或执行智能体函数。智能体函数是通过智能体程序在物理架构上的执行而产生的。

4. 智能体的类别

智能体有 5 种主要类别。基于其能力和感知智能水平来分组。

（1）简单反射型智能体

这些智能体使用当前感知而不是感知历史来执行动作。条件-动作规则被用作智能体函数的基础。在此类智能体中，完全可观察的环境是智能体函数成功的理想选择。

（2）基于模型的反射型智能体

与简单的反射型智能体不同，基于模型的反射型智能体在其行动中考虑感知历史。即使是在无法完全观察到的环境中，智能体函数仍然可以很好地工作。这些智能体使用内部模型来确定行动的感知历史和动作效果。它们反映了当前状态中的某些未被观察到的方面。

（3）基于目标的智能体

这些智能体比基于模型的反射型智能体的能力更强。基于目标的智能体使用目标信息来描述所需的能力。这允许它们在各种可能性中进行选择。这些智能体选择可以促进目标实现的最佳行动。

（4）基于效用的智能体

这些智能体根据效用做出选择。因为有附加的效用度量组件，它们比基于目标的智能体更先进。使用效用函数，将状态映射到某个效用度量。理性智能体会选择能够优化结果预期效用的操作。

（5）学习智能体

这些智能体能够从以前的经验中学习。学习智能体具有以下元素。

学习元素：该元素使学习智能体能够从以前的经验中学习。

评价：它反馈智能体的工作情况。

执行元素：该元素决定需要采取的外部行动。

问题生成器：它充当执行某些任务的反馈智能体，如提出（新的）建议和保存历史记录。

5. 智能体如何工作

智能体通过传感器、执行器和效应器三个主要组件工作。了解这些组件的概况可以加深我们对智能体如何工作的理解。

传感器：这些是检测环境变化的设备。它们收集的信息被发送到其他设备。在人工智能中，智能体通过传感器观察系统的环境。

执行器：这些是将能量转换为行动的组件。它们扮演控制和移动系统的角色。示例包括导轨、电机和齿轮。

效应器：环境受效应器的影响。示例包括腿、手指、轮子、显示屏和手臂。

图 10-1 显示了这些组件在 AI 系统中的定位。

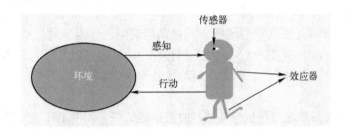

图 10-1　组件在 AI 系统中的定位

智能体通过传感器接收来自环境的输入（感知）。该智能体使用人工智能，根据获得的信息/观察结果做出决策。然后通过执行器启动行动。如图 10-2 所示，未来的决策将受到感知历史和过去行动的影响。

6. 智能体的应用

人工智能中的智能体已应用于许多现实生活环境中。

（1）信息搜索、检索和导航

智能体增强了信息的访问和导航。这是通过使用搜索引擎搜索信息来实现的。互联网由许多数据对象组成，用户可能需要花费大量时间来搜索特定的数据对象。智能体代表用户在短时间内执行此任务。

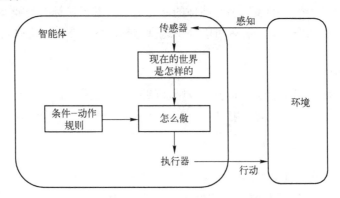

图 10-2　智能体工作流程

（2）重复办公活动

一些公司已将某些管理任务自动化以降低运营成本。一些已自动化的功能领域包括客户支持和销售。智能体也被用于提高办公效率。

（3）医学诊断

智能体也已应用于医疗保健服务，以改善患者的健康。在这种情况下，患者被视为环境。计算机键盘用作接收患者症状数据的传感器。智能体使用此信息来决定最佳行动方案。医疗护理通过执行器提供，如测试和治疗。

（4）环境清洁

智能体还用于提高环境清洁的效率和清洁度。在这种情况下，环境可以是房间、桌子或地毯。环境清洁中使用的一些传感器包括摄像头、碰撞传感器和污垢检测传感器。行动由执行器（如刷子、轮子和真空吸尘器）启动。

（5）自动驾驶

智能体增强了自动驾驶汽车的运行。在自动驾驶中，使用各种传感器从环境中收集信息。其中包括摄像头、GPS 和雷达。在此应用程序中，环境可以是行人、其他车辆、道路或路标。各种执行器用于启动动作。例如，制动器用于使汽车停止。

7. 结论

智能体通过代表系统或用户执行某些耗时且困难的任务，使工作更轻松。这些智能体使某些任务的自动化成为可能。

随着技术的不断进步，智能体的发展也会得到加强。智能体将进一步转化为复杂的人工智能驱动设备，以应对当前的全球挑战。这项有趣的技术似乎无往不胜。

Text B

Smart Manufacturing

扫码听课文

Smart manufacturing is a powerful disruptive force with the potential to restructure the current competitive landscape and produce a new set of market leaders. Companies that are slow to adopt new technologies and processes could be left behind.

1. What is smart manufacturing

Smart manufacturing is a broad concept. It is not something that can be implemented in a production process directly. It is a combination of various technologies and solutions which collectively, if implemented in a manufacturing ecosystem, is termed smart manufacturing. We call these technologies and solutions "enablers," which help in optimizing the entire manufacturing process and thus increase overall profits.

Some of the prominent enablers in the current market scenario include artificial intelligence, blockchain in manufacturing, Industrial Internet of Things (IIoT), robotics, condition monitoring and cyber security.

Companies are constantly investing and exploring how to obtain benefits through the implementation of enablers. If we closely look at enablers, then we will observe that they are either generating data or accepting data, or both. Data analysis will help in making the production process efficient, transparent and flexible.

2. How data drives smart manufacturing

Smart manufacturing is all about harnessing data, data will tell us "what to do" and "when to do it." Since smart factories are built around data, cyber security, above all, will play an important and significant role in the entire ecosystem of smart manufacturing. Data security is an important challenge while implementing these enablers.

All the stakeholders of smart manufacturing may be characterized typically in three types of companies, which can broadly be called "product and control solution providers" "IT solution providers or enablers" and "connectivity solution providers."

- Product and control solution providers include all the companies involved in the development of automation products and services. Examples include ABB, Honeywell, Siemens, Emerson, Rockwell, Yokogawa and Schneider.
- IT solution providers or enablers power the whole concept of IIoT and asset management. They help in building control, monitoring and analytics infrastructures. Examples include HP, Microsoft, SAS, Oracle and Intel.
- Connectivity solution providers are telecom service providers that facilitate the smooth flow of data for asset management. Examples include Cisco, Huawei and AT&T.

3. The implementation of industrial internet of things

Industrial internet of things is nothing but an ecosystem where every device, machine or process is connected through data communication systems. Each machine and piece of industrial equipment is embedded or connected with sensors which typically generate the relevant data. This is further transferred to the cloud/software systems through data communication systems. This huge amount of data has lots of insight which if analysed may help in identifying certain dark areas within the production process. After the analysis of the data, it is sent as feedback to the production systems for any corrective action.

There is huge potential for IIoT in smart manufacturing. You can't increase production beyond certain limits, so what do you do to increase your profits? You can't increase production because there is no demand for that. So, you try to look at the backend process and make it efficient. Now this is possible only when you have the precise details about your production process. This is where IIoT comes into the picture. Sensors generating data can be implemented at each process of production so that you can get the data, analyse it and take corrective action to increase the efficiency, thus increasing profitability.

However, it is not so easy to implement IIoT in current and old organizations, but you can implement it in newly established manufacturing facilities. This is because results can only be achieved if the implementation of smart manufacturing concept is there right from the start of design process for a manufacturing facility.

4. The rise of artificial intelligence in manufacturing

The concept of artificial intelligence is old, but it is now finding applications in manufacturing ecosystems. In the last 5-6 years, there has been a tremendous increase in interest and investment regarding AI in manufacturing. This is mainly due to a few reasons, as AI will work only if the data is available, and it has only recently been possible to build the needed capability to generate huge amount of data with low-cost sensors, store data in low-cost systems, process data at affordable rates.

These have collectively made it possible for AI to be implemented in manufacturing shop floors. Earlier manufacturing was being done by low-cost countries where it is very difficult to justify the high cost of implementation of AI in their manufacturing ecosystems. But due to a rise in wages, now it is possible to implement AI in many countries.

Furthermore, robots with AI capability are also finding significant application in manufacturing ecosystems. Robotics with AI capability has made it possible to involve the perception-based decision making. Predictive maintenance is another important aspect where AI finds significant application. Predictive maintenance enables the capability to determine performance, breakdown and operating conditions of equipment or machine on a real-time basis.

In manufacturing, the extensive usage of big data technology, industrial IoT, robotics computer vision technology, cross-industry partnerships and collaborations, and significant increase in venture capital investments will propel the growth of the AI in manufacturing market.

5. The future of blockchain in manufacturing

Blockchain in manufacturing is still at a very nascent stage. However, it is a much-discussed new technology in manufacturing ecosystems. Currently, it is being implemented in financial systems, but companies are exploring its application in manufacturing.

Looking at the capabilities of blockchain, aviation, food and beverage, and medical are some of the industries which could greatly benefit from this technology. These industries, due to some stringent rules and regulations, require full scrutiny of all their suppliers across the value chain. Blockchain could help in maintaining quality control right from the development of raw materials. Currently, most of the attention is on the development of blockchain for supply chain function across the manufacturing ecosystems.

Some of the industries that are actively developing blockchain include apparel, solar energy, mining, fishing, food and beverage, shipping (cargo transportation), fertilizer, healthcare and aviation. The list is not exhaustive, and as the technology matures, more and more industries may get involved in implementing blockchain. Companies such as IBM, Microsoft, GE, Samsung, and Moog are involved in developing and implementing the blockchain in manufacturing ecosystems.

The blockchain market in manufacturing has yet to conceptualize fully, and thus we are expecting the market to start generating significant revenue from 2023 onwards. However, many organizations have already started investing and exploring the benefit from blockchain technology in manufacturing ecosystems.

6. The importance of industrial robotics

The next thing that makes a typical manufacturing plant a smart manufacturing facility is the implementation of industrial robots. Industrial robots is not a new concept. It has been in the systems for the last 40-50 years. The only thing that has changed with respect to industrial robots is that they have now become intelligent. Earlier the robots were programmed to do one single task at a time. If you want to do other type of tasks, then you must change the codes.

Now robots are well connected with the sensor network implemented within the manufacturing shop floor, and they get the data from sensors and change their action accordingly. Artificial intelligence is also being slowly implemented in robotics systems, and thus it makes systems autonomous. Through AI, robotics systems are expected to change their actions according to the situation on a real-time basis.

The main drivers for the industrial robotics market are increasing investments for automation in various industries and the growing demand from Small and Medium-scale Enterprises (SME) in developing countries.

7. The benefits of digital twin

Digital twin is another concept in the ecosystems of smart manufacturing. It creates a virtual model by using the data obtained from sensors in the systems or asset. Predictive maintenance is one of the important systems which will use digital twin. The benefits of digital twin include potential reduction in time and cost of product development and elimination of unplanned downtime. The rising adoption of IoT and cloud platforms, and 3D printing and 3D simulation software are boosting the adoption of digital twin.

Aerospace and defense, automotive and transportation, electronics and electrical/machine manufacturing, and energy and utility are the major adopter of digital twin. Once the concept of digital twins develops and matures, we may see its increasing application in non-manufacturing sectors such as retail and the consumer goods market.

✍ New Words

disruptive	[dɪs'rʌptɪv]	*adj.*颠覆性的，破坏性的
potential	[pə'tenʃl]	*n.*潜力；可能性
		*adj.*潜在的

restructure	[ˌriːˈstrʌktʃə]	v.重构，重建，重组
collectively	[kəˈlektɪvli]	adv.全体地，共同地
ecosystem	[ˈiːkəʊsɪstəm]	n.生态系统
enabler	[ɪˈneɪblə]	n.推动者
profit	[ˈprɒfɪt]	n.利润；好处 v.获益；得益于
blockchain	[ˈblɒktʃeɪn]	n.区块链
robotics	[rəʊˈbɒtɪks]	n.机器人技术
transparent	[trænsˈpærənt]	adj.透明的
flexible	[ˈfleksəbl]	adj.灵活的
harness	[ˈhɑːnɪs]	vt.利用；控制
connectivity	[kəˌnekˈtɪvəti]	n.连通性
telecom	[ˈtelɪˌkɒm]	n.电信
embed	[ɪmˈbed]	v.（使）嵌入，融入
backend	[bækend]	n.后端
precise	[prɪˈsaɪs]	adj.清晰的；精确的
corrective	[kəˈrektɪv]	adj.改正的；矫正的
tremendous	[trəˈmendəs]	adj.极大的，巨大的
low-cost	[ˌləʊ ˈkɒst]	adj.低成本的，价格便宜的
affordable	[əˈfɔːdəbl]	adj.价格合理的
propel	[prəˈpel]	vt.推进；推动
stringent	[ˈstrɪndʒənt]	adj.严格的
regulation	[ˌregjuˈleɪʃn]	n.规章，规则 adj.规定的
supplier	[səˈplaɪə]	n.供应商，供应者
basis	[ˈbeɪsɪs]	n.基础；基准，根据；方式
unplanned	[ˌʌnˈplænd]	adj.无计划的，未筹划的

✍ Phrases

smart manufacturing	智能制造
condition monitoring	状态监测
cyber security	网络安全
smart factory	智能工厂
significant role	重要作用
asset management	资产管理
data communication system	数据通信系统

dark area	盲点，暗区
design process	设计过程
predictive maintenance	预测性维护
venture capital	风险投资，风险资本
industrial robot	工业机器人
with respect to	关于，至于，谈到
digital twin	数字孪生
virtual model	虚拟模型
consumer good	消费品

✍ Abbreviations

IIoT (Industrial Internet of Thing)	工业物联网

Reference Translation

智能制造

智能制造是一股强大的颠覆性力量，有可能重构当前的竞争格局并产生一批新的市场领导者。较晚采用新技术和流程的公司可能会被甩在后面。

1. 什么是智能制造

智能制造是一个广义的概念。它不是可以直接在生产过程中实施的东西。它是各种技术和解决方案的组合，如果在制造生态系统中实施，则统称为智能制造。我们将这些技术和解决方案称为"推动因素"，它们有助于优化整个制造过程，从而提高整体利润。

当前市场情景中的一些突出推动因素包括人工智能、制造业中的区块链、工业物联网（IIoT）、机器人、状态监测、网络安全。

企业在不断地投资和探索如何通过实施推动因素来获取收益。如果仔细观察推动因素，我们会发现它们要么生成数据，要么接收数据，或两者兼有。数据分析将有助于使生产过程高效、透明和灵活。

2. 数据如何驱动智能制造

智能制造就是利用数据，数据会告诉我们"做什么"和"什么时候做"。由于智能工厂是围绕数据构建的，因此网络安全首先将在整个智能制造生态系统中发挥重要作用。在实施这些推动因素时，数据安全是一项重要挑战。

智能制造的所有利益相关者通常可以分为三类，广义上可以称为"产品和控制解决方案提供商""IT 解决方案提供商或推动者"和"连接解决方案提供商"。

- 产品和控制解决方案供应商包括所有参与自动化产品和服务开发的公司。例如，ABB、霍尼韦尔、西门子、艾默生、罗克韦尔、横河和施耐德。
- IT 解决方案提供商或推动者推动了工业物联网和资产管理的整个概念。它们有助于构建控制、监控和分析基础设施。例如，HP、Microsoft、SAS、Oracle 和 Intel。

- 连接解决方案提供商是电信服务提供商，可促进资产管理数据的顺畅流动。例如，思科、华为和 AT&T。

3. 工业物联网的实施

工业物联网只不过是一个生态系统，其中每个设备、机器或过程都通过数据通信系统进行连接。每台机器和工业设备都嵌入或连接有传感器，这些传感器通常会生成相关数据。这些数据通过数据通信系统进一步传输到云/软件系统。这些海量数据有很多洞察力，如果对其进行分析，可能有助于识别生产过程中的某些盲点。对数据进行分析后，将其作为反馈发送给生产系统，以便采取任何纠正措施。

工业物联网在智能制造领域具有巨大潜力。如果不能无限扩大生产，那么如何增加利润？因为没有需求，所以你不能增加产量。因此，你尝试查看后端流程并使其高效。现在，只有当你掌握了有关生产过程的精确细节时，这才可能。这就是工业物联网的用武之地。可以在每个生产过程中安装生成数据的传感器，以便获取数据、分析数据并采取纠正措施来提高效率，从而提高盈利能力。

然而，在当前和旧的组织中实施工业物联网并非易事，但可以在新建立的制造设施中实施它。这是因为只有从制造设施的设计过程开始就实施智能制造，才能取得成果。

4. 制造业中人工智能的兴起

人工智能的概念由来已久，但现在正在制造业生态系统中得到应用。在过去的 5～6 年中，人们对人工智能在制造业的兴趣和投资大幅增加。这主要是由于几个原因，因为人工智能只有在数据可用的情况下才能工作，而且直到最近才有可能构建所需的能力，从而使用低成本传感器生成大量数据、将数据存储在低成本系统中、以可承受的价格处理数据。

所有这些因素使人工智能在制造车间实施成为可能。早期的制造是由低成本国家完成的，在这些国家，很难证明在其制造生态系统中实施人工智能的高成本是合理的。但由于人工工资上涨，现在许多国家都有意实施人工智能。

此外，具有人工智能能力的机器人在制造生态系统中也有着重要的应用。具有人工智能能力的机器人技术使基于感知的决策成为可能。预测性维护是人工智能的另一个重要应用方面。预测性维护能够实时确定设备或机器的性能、故障和运行状况。

在制造业，大数据、工业物联网、机器人和计算机视觉技术的广泛使用、跨行业伙伴关系与协作以及风险投资的显著增加，将推动人工智能在制造业市场的增长。

5. 制造业中区块链的未来

制造业中的区块链仍处于非常初级的阶段。然而，它是制造业生态系统中一项被广泛讨论的新技术。目前，区块链正在金融系统中实施，但很多公司正在探索其在制造业中的应用。

看看区块链的能力，航空、食品和饮料以及医疗是一些可以从这项技术中受益匪浅的行业。由于一些严格的规则和法规，这些行业需要对整个价值链中的所有供应商进行全面审查。区块链可以帮助从原材料的开发开始就进行质量控制。目前，大部分注意力都集中在为整个制造生态系统的供应链功能开发区块链上。

一些积极开发区块链的行业包括服装、太阳能、采矿、渔业、食品和饮料、航运（货物运输）、化肥、医疗保健和航空。这个清单并不详尽，随着技术的成熟，可能会有越来越多的行业参与实施区块链。IBM、微软、GE、三星和穆格等公司都参与了在制造生态系统中开发和实施区块链的工作。

　　制造业中的区块链市场尚未完全概念化，因此我们预计该市场将从 2023 年开始产生可观的收入。然而，许多组织已经开始投资和探索区块链技术在制造业生态系统中的好处。

6. 工业机器人的重要性

　　使典型的制造工厂成为智能制造设施的下一件事是工业机器人的实施。工业机器人并不是一个新概念。它已经在系统中使用了 40～50 年。工业机器人的唯一改变是它们现在变得智能化了。早些时候，机器人被编程为一次完成一项任务。如果想执行其他类型的任务，则必须更改程序代码。

　　现在，机器人与制造车间内实施的传感器网络连接良好，它们从传感器获取数据并相应地改变它们的动作。人工智能也慢慢地在机器人系统中实施，因此它使系统具有自主性。通过人工智能，机器人系统有望根据情况实时改变其动作。

　　工业机器人市场的主要驱动力是各个行业对自动化的投资不断增加，以及发展中国家的中小型企业（SME）的需求不断增长。

7. 数字孪生的好处

　　数字孪生是智能制造生态系统中的另一个概念。它使用从系统或资产中的传感器获取的数据创建虚拟模型。预测性维护是使用数字孪生的重要系统之一。数字孪生的好处包括可能减少产品开发的时间和成本以及消除计划外停机。物联网和云平台以及 3D 打印和 3D 模拟软件的日益普及正在推动数字孪生的应用。

　　航空航天与国防、汽车与运输、电子与电气/机器制造以及能源与公用事业是数字孪生的主要采用者。一旦数字孪生的概念发展成熟，我们可能会看到它在零售和消费品市场等非制造业领域越来越多的应用。

Exercises

[Ex. 1] Answer the following questions according to Text A.

　　1. What is an Intelligent Agent (IA) ? What can it be described as?

　　2. What do the two main functions of intelligent agents include? What do intelligent agents consist of ?

　　3. How many main parts does the structure of IA consist of ? What are they?

　　4. What are the 5 main categories of intelligent agents? What is the grouping of these agents based on?

　　5. How do utility-based agents make choices? Why are they more advanced than goal-based agents?

　　6. What elements do learning agents have?

　　7. How do intelligent agents work?

　　8. What are sensors and actuators?

　　9. What are the many real-life situations intelligent agents in artificial intelligence have been applied in?

10. How have intelligent agents been applied in healthcare services to improve the health of patients?

[Ex. 2] Fill in the following blanks according to Text B.

1. Smart manufacturing is a combination of _____ and _____which collectively, if implemented in a manufacturing ecosystem, is termed _____.

2. Some of the prominent enablers in the current market scenario include _____, _____, _____, _____, condition monitoring and Cyber security.

3. All the stakeholders of smart manufacturing may be characterized typically in three types of companies, which can broadly be called _____, _____ and _____.

4. IT solution providers or enablers power the whole concept of IIoT and asset management. They help in _____, _____ and _____. Examples include HP, BM, _____, SAS, Oracle and Intel.

5. Industrial internet of things is nothing but _____ where every device, machine or process is connected through _____. Each machine and piece of industrial equipment is _____ or connected with _____ which typically _____.

6. Sensors generating data can be implemented at each process of _____ so that you can get the data, _____ and take corrective action to _____, thus increasing _____.

7. Predictive maintenance enables the capability to _____, _____ and _____ of equipment or machine _____.

8. Looking at the capabilities of blockchain, _____, _____, and medical are some of the industries which could greatly benefit from this technology. These industries, due to some _____ and _____, require full scrutiny of all their suppliers across the value chain.

9. Now robots are well connected with _____ implemented within the manufacturing shop floor, and they _____ from sensors and _____.

10. The benefits of digital twin include potential reduction in time and cost of _____ and _____. Aerospace and defense, _____, electronics as electrical/machine manufacturing, and _____ are the major adopter of digital twin.

[Ex. 3] Translate the following terms or phrases from English into Chinese.

1. utility	1.	
2. blockchain	2.	
3. ecosystem	3.	
4. embed	4.	
5. transparent	5.	
6. autonomous driving	6.	
7. AI-driven device	7.	

8. data object   8. _____

9. operating cost   9. _____

10. cyber security   10. _____

[Ex. 4] Translate the following terms or phrases from Chinese into English.

1. *n.*设备，装置   1. _____

2. *v.*执行，实施   2. _____

3. *n.*反馈   3. _____

4. *vt.*使最优化   4. _____

5. *n.*响应，反应   5. _____

6. 工业机器人   6. _____

7. 预测性维护   7. _____

8. 智能工厂   8. _____

9. 风险投资，风险资本   9. _____

10. 虚拟模型   10. _____

[Ex. 5] Translate the following passage into Chinese.

Smart City

Emerging trends such as automation, machine learning and the IoT are driving smart city adoption.

Theoretically, any area of city management can be incorporated into a smart city initiative. A classic example is the smart parking meter that uses an application to help drivers find available parking spaces without prolonged circling of crowded city blocks. The smart meter also enables digital payment, so there's no risk of coming up short of coins for the meter.

Also in the transportation arena, smart traffic management is used to monitor and analyze traffic flows in order to optimize streetlights and prevent roadways from becoming too congested based on time of day or rush-hour schedules. Smart public transit is another facet of smart cities. Smart transit companies are able to coordinate services and fulfill riders' needs in real time, improving efficiency and rider satisfaction. Ride-sharing and bike-sharing are also common services in a smart city.

Energy conservation and efficiency are major focuses of smart cities. Using smart sensors, smart streetlights dim when there aren't cars or pedestrians on the roadways. Smart grid technology can be used to improve operations, maintenance and planning, and to supply power on demand and monitor energy outages.

Smart city initiatives also aim to monitor and address environmental concerns such as climate change and air pollution. Waste management and sanitation can also be improved with smart technology, be it using internet-connected trash cans and IoT-enabled fleet management systems for waste collection and removal, or using sensors to measure water parameters and guarantee the quality of drinking water at the front end of the system, with proper wastewater removal and drainage at the back end.

Smart city technology is increasingly being used to improve public safety. For example, smart sensors can be critical components of an early warning system before droughts, floods, landslides or hurricanes.

Smart buildings are also often part of a smart city project. Legacy infrastructure can be retrofitted and new buildings constructed with sensors to not only provide real time space management and ensure public safety, but also to monitor the structural health of buildings. Sensors can detect wear and tear, and notify officials when repairs are needed. Citizens can help in this matter, notifying officials through a smart city application when repairs are needed in buildings and other public infrastructure, such as potholes. Sensors can also be used to detect leaks in water mains and other pipe systems, helping reduce costs and improve the efficiency of public workers.

Smart city technologies also bring efficiencies to urban manufacturing and urban farming, including job creation, energy efficiency, space management and fresher goods for consumers.

Unit 11

Text A

Five Programming Languages for AI

扫码听课文

AI projects differ from traditional software projects. The differences lie in the technology stack, the skills required for an AI-based project, and the necessity of deep research. To implement your AI aspirations, you should use a programming language that is stable, flexible and has tools available.

AI is still developing and growing, and there are several languages that dominate the development landscape. Here we offer a list of programming languages that provide ecosystems for developers to build projects with AI.

1. Python

From development to deployment and maintenance, Python helps developers be productive and confident about the software they're building. Benefits that make Python the best fit for AI-based projects include simple and consistent, extensive selection of libraries and frameworks, platform independence, and great community and popularity. These add to the overall popularity of the language.

The reasons why Python is used for AI are as follows.

(1) Simple and consistent

Python offers concise and readable code. While complex algorithms and versatile workflows stand behind AI, Python's simplicity allows developers to write reliable systems. Developers can put all their effort into solving a ML problem instead of focusing on the technical nuances of the language. Additionally, Python is appealing to many developers as it's easy to learn. Python code is understandable by humans, which makes it easier to build models for machine learning.

Many programmers say that Python is more intuitive than other programming languages. Others point out the many frameworks, libraries and extensions that simplify the implementation of different functionalities. It's generally accepted that Python is suitable for collaborative implementation when multiple developers are involved. Since Python is a general-purpose language, it can do a set of complex machine learning tasks and enable you to build prototypes quickly that allow you to test your product for machine learning purposes.

(2) Extensive selection of libraries and frameworks

Implementing AI algorithms can be tricky and requires a lot of time. It's vital to have a well-structured and well-tested environment to enable developers to come up with the best coding solutions.

To reduce development time, programmers turn to a number of Python frameworks and libraries. A software library is pre-written code that developers use to solve common programming tasks. Python, with its rich technology stack, has an extensive set of libraries for artificial intelligence. Some of them are Keras, TensorFlow and Scikit-learn for machine learning; NumPy for high-performance scientific computing and data analysis; SciPy for advanced computing; Pandas for general-purpose data analysis; and Seaborn for data visualization.

(3) Platform independence

Platform independence refers to a programming language or framework allowing developers to implement things on one machine and use them on another machine without any (or with only minimal) changes. One key to Python's popularity is that it's a platform independent language. Python is supported by many platforms including Linux, Windows and macOS. Python code can be used to create standalone executable programs for most common operating systems, which means that Python software can be easily distributed and used on those operating systems without a Python interpreter.

What's more, developers usually use services such as Google or Amazon for their computing needs. However, you can often find companies and data scientists using their own machines with powerful Graphics Processing Units (GPU) to train their ML models. And the fact that Python is platform independent makes this training a lot cheaper and easier.

(4) Great community and popularity

Python is one of the most popular programming languages, which ultimately means that you can find and hire a development company with the necessary skill set to build your AI-based project.

Online repositories contain over 140,000 custom-built Python software packages. Scientific Python packages such as Numpy, Scipy and Matplotlib can be installed in a program running on Python. These packages cater to machine learning and help developers detect patterns in big sets of data. Python is so reliable that Google uses it for crawling web pages, Pixar uses it for producing movies, and Spotify uses it for recommending songs.

It's a well-known fact that the Python AI community has grown across the globe. There are Python forums and an active exchange of experience related to machine learning solutions. For any task you may have, the chance is pretty high that someone else out there has dealt with the same problem. You can find advice and guidance from developers. You are sure to find the best solution to your specific needs if you turn to the Python community.

2. R

R is generally applied when you need to analyze and manipulate data for statistical purposes. R has packages such as Gmodels, Class, Tm and RODBC that are commonly used for building machine learning projects. These packages allow developers to implement machine learning algorithms without extra hassle and let them quickly implement business logic.

R was created by statisticians to meet their needs. This language can give you in-depth statistical analysis whether you're handling data from an IoT device or analyzing financial models.

What's more, if your task requires high-quality graphs and charts, you may want to use R. With ggplot2, ggvis, GoogleVis, Shiny, rCharts and other packages, R's capabilities are greatly extended, helping you turn visuals into interactive web apps.

Compared with Python, R has a reputation for being slow and lagging when it comes to large-scale data products. It's better to use Python or Java, with its flexibility for actual product development.

3. Scala

Scala is invaluable when it comes to big data. It offers data scientists an array of tools such as Saddle, ScalaLab and Breeze. Scala has great concurrency support, which helps with processing large amounts of data. Since Scala runs on the JVM, it goes beyond all limits hand in hand with Hadoop, an open source distributed processing framework that manages data processing and storage for big data applications running in clustered systems. Scala is highly maintainable.

4. Julia

If you need to build a solution for high-performance computing and analysis, you might want to consider Julia. Julia has a similar syntax to Python and was designed to handle numerical computing tasks. Julia provides support for deep learning via the TensorFlow.jl wrapper and the Mocha framework.

However, the language is not supported by many libraries and doesn't yet have a strong community like Python because it's relatively new.

5. Java

Another language worth mentioning is Java. Java is object-oriented, portable, maintainable and transparent. It's supported by numerous libraries such as WEKA and RapidMiner.

Java is widespread when it comes to natural language processing, search algorithms and neural networks. It allows you to quickly build large-scale systems with excellent performance.

But if you want to perform statistical modeling and visualization, then Java is the last language you want to use. Even though there are some Java packages that support statistical modeling and visualization, they aren't sufficient.Moreover, Python has advanced tools that are well supported by the community.

New Words

project	['prɒdʒekt]	n.项目，工程
stack	[stæk]	n.堆栈
aspiration	[ˌæspə'reɪʃn]	n.抱负，志向
stable	['steɪbl]	adj.稳定的；牢固的
deployment	[dɪ'plɔɪmənt]	n.部署
confident	['kɒnfɪdənt]	adj.坚信的；自信的
simplicity	[sɪm'plɪsəti]	n.简单，朴素
consistency	[kən'sɪstənsi]	n.一致性，连贯性
library	['laɪbrəri]	n.库
community	[kə'mju:nəti]	n.社区，社团

popularity	[ˌpɒpjuˈlærəti]	n.流行性，普及度
consistent	[kənˈsɪstənt]	adj.一致的；连贯的
concise	[kənˈsaɪs]	adj.简明的，简洁的
readable	[ˈriːdəbl]	adj.易读的；易懂的，可读的
workflow	[ˈwɜːkfləʊ]	n.工作流程
nuance	[ˈnjuːɑːns]	n.细微差别
appeal	[əˈpiːl]	v.对……有吸引力
		n.吸引力
programmer	[ˈprəʊgræmə]	n.程序设计者，程序员
intuitive	[ɪnˈtjuːɪtɪv]	adj.直观的；直觉的
functionality	[ˌfʌŋkʃəˈnæləti]	n.功能；功能性
collaborative	[kəˈlæbərətɪv]	adj.合作的，协作的
extensive	[ɪkˈstensɪv]	adj.广阔的；广泛的；大量的
tricky	[ˈtrɪki]	adj.难办的，棘手的
visualization	[ˌvɪʒʊəlaɪˈzeɪʃn]	n.可视化
standalone	[ˈstændəˌləʊn]	adj.单独的，独立的
interpreter	[ɪnˈtɜːprɪtə]	n.解释器，解释程序
online	[ˌɒnˈlaɪn]	adj.在线的；联网的；联机的
repository	[rɪˈpɒzətri]	n.仓库
detect	[dɪˈtekt]	v.发现；查明；测出
crawl	[krɔːl]	v.爬行
forum	[ˈfɔːrəm]	n.论坛
experience	[ɪkˈspɪəriəns]	n.经验；经历
		v.经历；感受
guidance	[ˈgaɪdns]	n.指导
dominate	[ˈdɒmɪneɪt]	v.控制，支配
hassle	[ˈhæsl]	n.困难的事情；麻烦的事情
interactive	[ˌɪntərˈæktɪv]	adj.交互式的；互动的
reputation	[ˌrepjuˈteɪʃn]	n.名誉，名声
lag	[læg]	vi.走得极慢，落后
invaluable	[ɪnˈvæljuəbl]	adj.非常宝贵的；无价的
concurrency	[kənˈkʌrənsi]	n.并发（性）
distributed	[dɪsˈtrɪbjuːtɪd]	adj.分布式的
maintainable	[meɪnˈteɪnəbl]	adj.可维护的
syntax	[ˈsɪntæks]	n.语法
object-oriented	[ˈɒbdʒɪkt ɔːˈriəntɪd]	adj.面向对象的
portable	[ˈpɔːtəbl]	adj.轻便的，可移植的
excellent	[ˈeksələnt]	adj.极好的，优秀的
sufficient	[səˈfɪʃnt]	adj.足够的，充足的，充分的

✎ Phrases

technology stack	技术堆栈
platform independence	平台独立性

build model	构建模型
point out	指出
be suitable for	适用于
general-purpose language	通用语言
come up with	想出，提供，提出；赶上
turn to	向……求助
pre-written code	预先编写的代码
high-performance scientific computing	高性能科学计算
data visualization	数据可视化
software package	软件包，程序包
business logic	业务逻辑
concurrency support	并发支持
hand in hand	手拉手，携手；密切合作
open source	开源
clustered system	集群系统
numerical computing	数值计算

 **Abbreviations**

GPU (Graphics Processing Unit)	图形处理单元
JVM (Java Virtual Machine)	Java 虚拟机

Reference Translation

用于人工智能的五种编程语言

人工智能项目不同于传统的软件项目。不同之处在于技术堆栈、基于 AI 的项目所需的技能以及深入研究的必要性。为了实现你的 AI 愿望，你应该使用一种稳定、灵活且有可用工具的编程语言。

人工智能仍在发展和成长，有几种语言主导着发展前景。在这里，我们提供一个编程语言列表，这些语言为开发人员使用 AI 构建项目提供了生态系统。

1. Python

从开发到部署和维护，Python 可以帮助开发人员提高工作效率，并且使他们对正在构建的软件充满信心。Python 最适合基于 AI 的项目的优势在于简单一致、广泛可选的库和框架、平台独立性、强大的社区和人气。这些增加了该语言的整体流行度。

Python 用于 AI 的原因如下。

（1）简单一致

Python 提供简洁易读的代码。虽然人工智能背后有复杂的算法和多种工作流，但 Python 的简单性允许开发人员编写可靠的系统。开发人员可以将所有精力放在解决机器学习问题上，而不是专注于语言的技术的细微差别。此外，Python 易于学习，因此吸引了许多开发人员。Python 代码是人类可以理解的，这使得构建机器学习模型变得更加容易。

许多程序员认为 Python 比其他编程语言更直观。其他人指出许多框架、库和扩展简化了不同功能的实现。当涉及多个开发人员时，公认 Python 适用于协同工作。Python 是一种通用语言，它可以执行一组复杂的机器学习任务，并使你能够快速构建原型，从而允许你出于机器学习的目的测试产品。

（2）广泛可选的库和框架

实施 AI 算法可能很棘手，而且需要大量时间。拥有结构良好且经过良好测试的环境对于开发人员能够提出最佳编码解决方案至关重要。

为了减少开发时间，程序员借助于许多的 Python 框架和库。软件库是开发人员用来解决常见编程任务的预先编写的代码。Python 凭借其丰富的技术堆栈，拥有大量用于人工智能的库。例如，用于机器学习的 Keras、TensorFlow 和 Scikit-learn；用于高性能科学计算和数据分析的 NumPy；用于高级计算的 SciPy；用于通用数据分析的 Pandas；以及用于数据可视化的 Seaborn。

（3）平台独立性

平台独立性是指一种编程语言或框架，允许开发人员在一台机器上实现程序并在另一台机器上使用它们而无须任何（或仅进行最小）更改。Python 流行的一个关键原因为它是一种独立于平台的语言。许多平台都支持 Python，包括 Linux、Windows 和 macOS。Python 代码可用于为大多数常见操作系统创建独立的可执行程序，这意味着 Python 软件可以轻松发布并在这些操作系统上使用，而无须 Python 解释器。

更重要的是，开发人员通常使用谷歌或亚马逊等服务来满足他们的计算需求。但是，你经常会发现公司和数据科学家使用自己的具有强大的图形处理单元（GPU）的机器来训练他们的机器学习模型。Python 独立于平台的事实使这种训练更加廉价和容易。

（4）强大的社区和人气

Python 是最流行的编程语言之一，这最终意味着你可以找到并雇佣具有必要技能的开发公司来构建基于 AI 的项目。

在线存储库包含超过 140,000 个定制的 Python 软件包。像 NumPy、SciPy 和 Matplotlib 等 Python 科学包可以安装到在 Python 上运行的程序中。这些软件包能够满足机器学习的需求并帮助开发人员发现大数据集中的模式。Python 非常可靠，谷歌用它来抓取网页，皮克斯用它来制作电影，而 Spotify 用它来推荐歌曲。

众所周知，Python AI 社区已经在全球范围内发展壮大。那里有 Python 论坛，可以交流与机器学习解决方案相关的积极经验。对于你可能要完成的任何任务，很可能其他人已经处理过相同的问题。你可以从开发人员那里找到建议和指导。如果你求助于 Python 社区，一定会找到满足你特定需求的最佳解决方案。

2. R

当你需要出于统计目的的分析和操作数据时，通常会应用 R。R 具有 Gmodels、Class、Tm 和 RODBC 等包，它们通常用于构建机器学习项目。这些包允许开发人员轻松实现机器学习算法，并让他们快速实现业务逻辑。

R 是统计学家为了满足他们的需求而创建的。无论你是处理来自物联网设备的数据还是分析财务模型，这种语言都可以提供深入的统计分析。

更重要的是，如果你的任务需要高质量的图形和图表，你可能需要使用 R。借助

ggplot2、ggvis、googleVis、Shiny、rCharts 和其他软件包，R 的功能得到了极大的扩展，可以帮助你将视觉效果转变为交互式 Web 应用程序。

与 Python 相比，R 在大规模数据产品方面以缓慢和滞后而著称。最好使用 Python 或 Java，它们在实际产品开发中更具灵活性。

3. Scala

在大数据方面，Scala 可以说是无价之宝。它为数据科学家提供了一系列工具，如 Saddle、ScalaLab 和 Breeze。Scala 具有强大的并发支持，这有助于处理大量数据。由于 Scala 在 JVM 上运行，因此它与 Hadoop 携手超越了所有限制，Hadoop 是一个开源分布式处理框架，用于管理集群系统中运行的大数据应用程序的数据处理和存储。Scala 是高度可维护的。

4. Julia

如果你需要构建高性能计算和分析的解决方案，你可以考虑 Julia。Julia 的语法与 Python 相似，旨在处理数值计算任务。Julia 通过 TensorFlow.jl 包装器和 Mocha 框架为深度学习提供支持。

但是，许多库不支持该语言，并且还没有像 Python 那样强大的社区，因为它相对较新。

5. Java

另一种值得一提的语言是 Java。Java 是面向对象、可移植、可维护和透明的。它得到了 WEKA 和 RapidMiner 等众多库的支持。

Java 在自然语言处理、搜索算法和神经网络方面应用广泛。它可以让你快速构建具有出色性能的大型系统。

但是，如果你想执行统计建模和可视化，那么 Java 是你最不想使用的语言。尽管有一些 Java 包支持统计建模和可视化，但它们还不够。另外，在这个方面，Python 拥有深受社区支持的高级工具。

Text B

AI Platforms

扫码听课文

While AI has significant potential, executing AI software development projects can be hard. Developing an AI solution isn't a one-dimensional project either since you might need to use several ways to achieve your objectives. For example, you might need to use Machine Learning (ML), Natural Language Processing (NLP), vision, speech and several other AI capabilities.

When you undertake a complex project, you need to use the right toolset, therefore, a robust set of AI development tools are important. Ideally, a robust AI development platform should offer the following artificial intelligence tools.

- ML capabilities like deep learning, supervised algorithms, unsupervised algorithms, etc.
- NLP capabilities like classification, machine translation, etc.

- Expert systems.
- Automation.
- Vision capabilities like image recognition or computer vision.
- Speech capabilities like speech-to-text and text-to-speech.

1. Microsoft Azure AI platform

As a cloud platform, Microsoft Azure hardly needs an introduction. Azure has made significant progress with its capabilities, and the Microsoft Azure AI platform is a popular choice for AI development.

The Azure AI platform offers all key AI capabilities as follows.

- Machine learning.
- Vision capabilities like object recognition.
- Speech capabilities like speech recognition.
- Language capabilities like machine translation.
- Knowledge mining.

The ML capabilities of the Azure AI platform include the following toolkits.

- Azure ML, which is a Python-based automated ML service. Azure ML works with popular open-source AI frameworks like TensorFlow.
- Azure Databricks, which is an Apache Spark-based big data service that integrates with Azure ML.
- ONNX, which is an open source model format.

The Azure AI platform has knowledge mining capabilities, and you can unlock insights from documents, images and media using it. This includes the following.

- Azure search, which is a cloud search service with built-in AI.
- Form recogniser, which is an AI-powered extraction service to transform your documents and forms into usable data.

The Azure AI platform offers AI apps, and you can customize them and use them in your application. This includes Azure cognitive services, which offer a wide collection of domain-specific pre-trained AI models. The Azure cognitive services include AI models for the following.

- Vision.
- Speech.
- Language.

The Azure AI platform has a development environment for creating bots, and it has templates for bots. It is also very popular for developing new open-source machine learning algorithms and software solutions.

2. Google Cloud AI platform

Google is yet another cloud computing giant that offers its AI platform. The Google Cloud AI platform offers all the key AI capabilities as follows.

(1) Machine learning

With the Google Cloud AI platform, you can easily develop your ML project and deploy it to production. The Google AI platform provides an integrated toolchain for this, which expedites the development and deployment.

With this platform, you can build portable ML pipelines using Kubeflow, which is an open-source platform from Google. You can deploy your ML project either on-premise or in the cloud. Cloud storage and BigQuery are the prominent options to store your data. You can access popular AI frameworks like TensorFlow.

(2) Deep learning

The Google Cloud AI platform offers pre-configured Virtual Machines (VM) for creating deep learning applications. You can provision this VM quickly on the Google Cloud, and the deep learning VM image contains popular AI frameworks.

You can launch Google compute engine instances where TensorFlow, PyTorch, Scikit-learn and other popular AI frameworks are already installed.

(3) Natural language processing

The Google Cloud AI platform has NLP capabilities, and you can use it to find out the meaning and structure of the text. You can use the Google NLP capabilities to analyze text, and the Google NLP API helps with this.

(4) Speech

The Google Cloud AI platform has APIs for speech-to-text and text-to-speech capabilities. On the one hand, its speech-to-text API can help you to convert audio to text, and it uses neural network models and vast datasets for this. The speech-to-text API supports more than 100 languages and their variations. With its speech recognition capabilities, you can enable voice command-and-control features in your app and the app can transcribe audio.

On the other hand, the Google text-to-speech API enables you to create a natural-sounding speech from text. You can convert texts into audio files of popular formats like MP3 or linear16.

(5) Vision

Vision is another key capability of the Google Cloud AI platform, and you can use this to derive insights from your images. The Google Cloud AI platform offers its vision capabilities through REST and RPC APIs, and these APIs use pre-trained ML models.

Your app can detect objects and faces. Moreover, it can read printed and handwritten texts using these APIs.

3. IBM Watson

IBM, the technology giant, has advanced AI capabilities, and IBM Watson is quite popular. There are already IBM Watson AI solutions specifically tailored for several industries like healthcare, oil and gas, advertising, financial services, media, Internet of Things (IoT), etc.

A key advantage of IBM Watson is that developers can use this platform to build their AI applications. It's an open AI for any cloud environment, and it's pre-integrated and pre-trained on flexible information architecture. This will expedite the development and deployment of your AI application.

IBM Watson offers the following to expedite your AI app development.

● It has developer tools like SDKs and detailed documentation for them.

- You can integrate Watson Assistant to build AI-powered conversational interfaces into your app.
- With IBM Watson, you can get Watson Discovery. It's an AI-powered search technology, and it can help your app to retrieve information that resides in silos.
- IBM Watson has Natural Language Processing (NLP) capabilities, and it's known as Watson Natural Language Understanding (NLU). The IBM Watson developer platform includes this.

You can also make use of the IBM Watson speech-to-text capabilities when you build on the Watson developer platform.

IBM Watson developer resources can be useful for your AI app development team. There are SDKs for Swift, Ruby, Java, Python, Node.js, .NET, etc., therefore, you will likely find a suitable SDK for your project.

4. Infosys Nia

Infosys Nia is an AI platform that allows you to build AI-powered apps. It offers the following AI capabilities.

- Machine learning: Nia Advanced ML offers a broad range of ML algorithms that operate at speed and scale. It makes building high-performing ML models easier.
- Contracts analysis: Nia contracts analysis capability includes ML, semantic modeling and deep learning.
- Nia chatbot: You can build AI-powered chatbots with Nia, and your app can have access to the enterprise knowledge repository. The app can also automate actions through a conversational interface.
- Nia data: Your AI app can integrate Nia data, a robust analytics solution.

5. Dialogflow

Dialogflow uses Google's infrastructure, so it has all the computation power you need. It incorporates Google's ML capabilities. It runs on the Google Cloud platform, therefore, you should be able to scale your AI app easily.

Dialogflow lets you build voice and text-based conversational interface for your app. Your app can run on web and mobile. You can connect your users on Google Assistant, Amazon Alexa, Facebook Messenger, etc.

The key capabilities offered by Dialogflow are ML, NLP.

6. BigML

BigML is highly focused on ML, and its development platform offers powerful ML capabilities. It provides robust ML algorithms, both for supervised and unsupervised learning.

You can implement instant access to its ML platform using its REST API, and you can do that both on-premises and on the cloud. BigML offers interpretable and exportable ML models, and this is a key advantage.

BigML offers the following features.

- It's programmable and repeatable. You can use popular languages like Python, Node.js, Ruby, Java, Swift, etc. to code your app, and BigML supports them.

- BigML helps you to automate your predictive modeling tasks.
- Deployment is flexible since you can deploy your AI app both on-premises or on the cloud. BigML has smart infrastructure solutions that help in scaling your app.
- BigML has robust security and privacy features.

New Words

significant	[sɪɡ'nɪfɪkənt]	adj.重要的；显著的
Achieve	[ə'tʃi:v]	v.实现，达到
undertake	[ˌʌndə'teɪk]	v.承担；从事
toolset	['tu:lset]	n.成套工具，工具箱
format	['fɔ:mæt]	n.格式
		vt.使格式化
insight	['ɪnsaɪt]	n.洞察力
built-in	[ˌbɪlt 'ɪn]	adj.嵌入的；内置的
service	['sɜ:vɪs]	n.服务
transform	[træns'fɔ:m]	v.转化
customize	['kʌstəmaɪz]	vt.定制，定做；用户化
template	['templeɪt]	n.样板；模板
giant	['dʒaɪənt]	n.巨人；大公司
toolchain	['tu:ltʃeɪn]	n.工具链
deploy	[dɪ'plɔɪ]	v.部署
expedite	['ekspədaɪt]	vt.加快进展；迅速完成
pipeline	['paɪplaɪn]	n.管道；渠道
on-premise	[ɒn 'premɪs]	n.本机端，本地
prominent	['prɒmɪnənt]	adj.突出的，杰出的
option	['ɒpʃn]	n.选项，选择；选择权
pre-configured	[pri:kən'fɪɡəd]	adj.预先配置的
variation	[ˌveəri'eɪʃn]	n.变体；变化，变动
pre-trained	['pri:treɪnd]	adj.预先训练的
pre-integrated	[pri:'ɪntɪɡreɪtɪd]	adj.预先集成的
conversational	[ˌkɒnvə'seɪʃənl]	adj.会话的
interface	['ɪntəfeɪs]	n.界面；接口
instant	['ɪnstənt]	adj.即时的，立即的
exportable	[eks'pɔ:təbəl]	adj.可导出的，可输出的
programmable	['prəʊɡræməbl]	adj.可编程的
repeatable	[rɪ'pi:təbl]	adj.可重复的
infrastructure	['ɪnfrəstrʌktʃə]	n.基础设施，基础架构
privacy	['praɪvəsi]	n.隐私

Phrases

cloud platform	云平台
object recognition	物体识别

knowledge mining	知识挖掘
cloud search	云搜索
extraction service	提取服务
cloud computing	云计算
cloud storage	云存储
analyze text	分析文本
convert ... to	把……转换为……
technology giant	科技巨头
contracts analysis	合约分析
semantic modeling	语义建模
computation power	计算能力
instant access	即时访问

Abbreviations

ONNX (Open Neural Network Exchange)	开放神经网络交换
VM (Virtual Machine)	虚拟机
API (Application Programming Interface)	应用程序接口
SDK (Software Development Kit)	软件开发工具包
NLU (Natural Language Understanding)	自然语言理解

Reference Translation

人工智能平台

虽然人工智能具有巨大的潜力，但执行人工智能软件开发项目可能会很困难。开发人工智能解决方案也不是一个一维项目，因为你可能需要使用多种方法来实现目标。例如，你可能需要使用机器学习（ML）、自然语言处理（NLP）、视觉、语音和其他几种人工智能功能。

当你承担一个复杂的项目时，需要使用正确的工具集，因此，一套强大的人工智能开发工具很重要。理想情况下，一个强大的人工智能开发平台应该提供以下人工智能工具。

- 机器学习能力，如深度学习、监督算法、无监督算法等。
- 自然语言处理功能，如分类、机器翻译等。
- 专家系统。
- 自动化。
- 视觉功能，如图像识别或计算机视觉。
- 语音功能，如语音转文本和文本转语音。

1. 微软 Azure 人工智能平台

作为一个云平台，微软 Azure 几乎不需要介绍。Azure 在功能方面取得了重大进展，微软 Azure AI 平台是人工智能开发的热门选择。

Azure AI 平台提供所有关键的人工智能功能，具体如下。

- 机器学习（ML）。
- 视觉功能，如物体识别。

- 语音识别等语音功能。
- 机器翻译等语言能力。
- 知识挖掘。

Azure AI 平台的机器学习功能包括以下工具包。

- Azure ML，这是一种基于 Python 的自动化机器学习服务。Azure ML 可与流行的开源人工智能框架（如 TensorFlow）配合使用。
- Azure Databricks，这是一种与 Azure ML 集成的基于 Apache Spark 的大数据服务。
- ONNX，这是一种开源模型格式。

Azure AI 平台具有知识挖掘功能，你可以使用它从文档、图像和媒体中获得洞察力。它包括以下内容。

- Azure 搜索，这是一种内置人工智能的云搜索服务。
- 表单识别器，这是一种人工智能驱动的提取服务，可将你的文档和表单转换为可用数据。

Azure AI 平台提供人工智能应用程序，你可以定制它们并在应用程序中使用它们。该平台含有多种 Azure 认知服务，它们提供了大量特定领域的预训练人工智能模型。Azure 认知服务包括以下人工智能模型。

- 视觉。
- 语音。
- 语言。

Azure AI 平台有一个用于创建机器人的开发环境，它有机器人模板。它在开发新的开源机器学习算法和软件解决方案方面也很受欢迎。

2. 谷歌云人工智能平台

谷歌是另一家提供人工智能平台的云计算巨头。Google Cloud AI 平台提供所有关键的人工智能功能，具体如下。

（1）机器学习

借助 Google Cloud AI 平台，你可以轻松开发机器学习项目并将其部署到生产环境中。谷歌人工智能平台为此提供了一个集成的工具链，可以加快开发和部署。

借助此平台，你可以使用 Kubeflow 构建可移植的机器学习管道，Kubeflow 是 Google 的一个开源平台。你可以在本地或云端部署机器学习项目。云存储和 BigQuery 是存储数据的重要选项。你可以访问流行的人工智能框架，如 TensorFlow。

（2）深度学习

Google Cloud AI 平台提供预配置的虚拟机（VM），用于创建深度学习应用程序。你可以在 Google Cloud 上快速配置此虚拟机，并且深度学习虚拟机映像包含流行的人工智能框架。

你可以启动已安装 TensorFlow、PyTorch、Scikit-learn 和其他流行人工智能框架的 Google Compute Engine 实例。

（3）自然语言处理

Google Cloud AI 平台具有自然语言处理功能，你可以使用它来找出文本的含义和结构。你可以使用 Google 自然语言处理功能来分析文本，而 Google NLP API 对此有所帮助。

（4）语音

Google Cloud AI 平台具有用于语音转文本和文本转语音功能的 API。一方面，它的语音转文本 API 可以帮助你将音频转换为文本，并且为此使用了神经网络模型和大量数据集。语音转文本 API 支持 100 多种语言及其变体。凭借其语音识别功能，你可以在应用程序中启用语音命令和控制功能，并且该应用程序可以转录音频。

另一方面，Google 语音转文本 API 使你能够根据文本创建听起来自然的语音文件。你可以将文本转换为 MP3 或 Linear16 等流行格式的音频文件。

（5）视觉

视觉是 Google Cloud AI 平台的另一项关键功能，你可以使用它从图像中获取洞察力。Google Cloud AI 平台通过 REST 和 RPC API 提供其视觉功能，这些 API 使用预训练的机器学习模型。

你的应用程序可以检测物体和人脸。此外，它可以使用这些 API 读取印刷文本和手写文本。

3．IBM Watson

科技巨头 IBM 拥有先进的人工智能能力，IBM Watson 颇受青睐。已经有专门为医疗保健、石油和天然气、广告、金融服务、媒体、物联网（IoT）等多个行业量身定制的 IBM Watson AI 解决方案。

IBM Watson 的一个关键优势是开发人员可以使用这个平台来构建他们的人工智能应用程序。它是适用于任何云环境的开放式人工智能，并且在灵活的信息架构上进行了预集成和预训练。这将加快你的人工智能应用程序的开发和部署。

IBM Watson 提供以下功能来加速你的人工智能应用程序开发。

- 它具有 SDK 等开发工具和详细的文档。
- 你可以将 Watson Assistant 集成到你的应用程序中，以构建人工智能驱动的对话界面。
- 使用 IBM Watson，你可以获得 Watson Discovery。它是一种基于人工智能的搜索技术，可以帮助你的应用检索驻留在孤岛中的信息。
- IBM Watson 具有自然语言处理（NLP）功能，称为 Watson 自然语言理解（NLU）。这包含在 IBM Watson 开发人员平台中。

在 Watson 开发人员平台上构建应用程序时，你还可以利用 IBM Watson 语音转文本功能。

IBM Watson 开发人员资源对你的人工智能应用程序开发团队很有用。有适用于 Swift、Ruby、Java、Python、Node.js、.NET 等的 SDK，因此，你可以找到适合你的项目的 SDK。

4．Infosys Nia

Infosys Nia 是一个人工智能平台，可让你构建基于人工智能的应用程序。它提供以下人工智能功能。

- 机器学习：Nia Advanced ML 提供了一系列快速和大规模的机器学习算法。它使构建高性能机器学习模型变得更加容易。
- 合约分析：Nia 合约分析能力包括机器学习、语义建模和深度学习。
- Nia 聊天机器人：你可以使用 Nia 构建由人工智能驱动的聊天机器人，你的应用程序可以访问企业知识库。该应用程序还可以通过对话界面自动执行操作。

● Nia 数据：你的人工智能应用程序可以集成 Nia 数据，这是一种强大的分析解决方案。

5. Dialogflow

Dialogflow 使用 Google 的基础架构，因此它具有你需要的所有计算能力。它结合了 Google 的机器学习功能。它在 Google Cloud 平台上运行，因此，你应该能够轻松扩展你的人工智能应用程序。

Dialogflow 可让你为应用程序构建基于语音和文本的会话界面。你的应用程序可以在网络和移动设备上运行。你可以在 Google Assistant、Amazon Alexa、Facebook Messenger 等上连接你的用户。

Dialogflow 提供的关键功能是机器学习、自然语言处理。

6. BigML

BigML 高度关注机器学习，其开发平台提供强大的机器学习能力。它为监督和无监督学习提供了强大的机器学习算法。

你可以使用其 REST API 实现对其机器学习平台的即时访问，并且你可以在本地和云上执行此操作。BigML 提供可解释和可导出的机器学习模型，这是一个关键优势。

BigML 提供以下功能。

● 可编程且可重复。你可以使用 Python、Node.js、Ruby、Java、Swift 等流行语言来编写你的应用程序，BigML 支持这些语言。

● BigML 帮助你自动执行预测建模任务。

● 部署灵活，因为你可以在本地或云端部署人工智能应用程序。BigML 拥有有助于扩展应用程序的智能基础架构解决方案。

● BigML 具有强大的安全和隐私功能。

Exercises

[Ex. 1] Answer the following questions according to Text A.

1. What do benefits that make Python the best fit for AI-based projects include?

2. What can Python do?

3. What does platform independence refer to?

4. What does "Python is one of the most popular programming languages" mean?

5. What are in the Python AI community?

6. What is R? What does R have?

7. When is Scala invaluable? What does it offer to data scientists?

8. What is Hadoop?

9. When might you want to consider Julia?

10. What is Java? What is it supported by?

[Ex. 2] Fill in the following blanks according to Text B.

1. The Azure AI platform offers all key AI capabilities such as _____; vision capabilities like _____; speech capabilities like _____; language capabilities

like _____; _____.

2. The ML capabilities of the Azure AI platform include the following toolkits:_____, _____ and _____.

3. The Azure cognitive services include AI models for _____, _____ and _____.

4. With the Google Cloud AI platform, you can easily _____ and _____. The Google AI platform provides _____ for this, which expedites the development and deployment.

5. _____ has APIs for speech-to-text and text-to-speech capabilities. Its speech-to-text API can help you to _____, and it uses _____ and vast datasets for this. The speech-to-text API supports _____ languages and their variations.

6. There are already IBM Watson AI solutions specifically tailored for several industries like _____, _____, _____, financial services, media, Internet of Things (IoT), etc. A key advantage of IBM Watson is that developers can use this platform to _____ .

7. With IBM Watson, you can get _____. It's an AI-powered search technology, and it can help your app to _____ that resides in silos. IBM Watson has Natural Language Processing (NLP) capabilities, and it's known as _____. You can also make use of the IBM Watson speech-to-text capabilities when you build on _____.

8. Infosys Nia is an AI platform that allows you to _____. It offers the following AI capabilities: _____, _____, _____ and _____.

9. Dialogflow uses _____, so it has all the computation power you need. The key capabilities offered by Dialogflow are _____, _____ and _____.

10. BigML is highly focused on _____, and its development platform offers _____. It provides robust ML algorithms, both for _____ and _____ learning.

[Ex. 3] **Translate the following terms or phrases from English into Chinese.**

1. collaborative   1. _____
2. concurrency   2. _____
3. consistency   3. _____
4. deployment   4. _____
5. distributed   5. _____
6. data visualization   6. _____
7. numerical computing   7. _____
8. platform independence   8. _____
9. software package   9. _____
10. cloud computing   10. _____

[Ex. 4] Translate the following terms or phrases from Chinese into English.

1. *n.*功能；功能性	1. _____
2. *adj.*交互式的；互动的	2. _____
3. *n.*解释器，解释程序	3. _____
4. *adj.*可维护的	4. _____
5. *n.*可视化	5. _____
6. 构建模型	6. _____
7. 云存储	7. _____
8. 计算能力	8. _____
9. 知识挖掘	9. _____
10. 物体识别	10. _____

[Ex. 5] Translate the following passage into Chinese.

5 Most Popular AI Programming Languages

1. Python

This language is widely used by programmers because of its pure syntax and the logical, strictly grammatical construction of the program. Python is most often used in the field of machine learning and neural network creation.

Hundreds of available libraries make it possible to implement almost any project, whether it is a web service, a mobile app, machine learning or scientific computing.

Key features:

- Relatively fast development speed.
- Rich and diverse set of libraries and tools.
- Balance of low-level and high-level programming.
- It allows you to test algorithms without having to implement them.
- It is still being actively developed.

2. LISP

Created back in 1958, this high-level programming language is widely used for AI development. Flexible and expandable, it is well suited as a tool for solving specific tasks.

It is worth mentioning that it was LISP that was used to create the very first robots that could move around, turn on/off the lights and move objects.

Key features:

- It can easily adapts to the solutions for specific needs.
- It has macros for implementing different AI levels.
- Automatic garbage collection.
- It supports rapid prototyping.

3. Prolog

AI developers appreciate it for its high level of abstraction, built-in search engine, nondeterminism, etc. This is one of the few languages that represents the paradigm of declarative programming.

Prolog is a language for implementing algorithms with a large implicit search of options (logical inference, finding dependencies, searching moves), which makes it a good choice for creating artificial intelligence.

Key features:

- Powerful and flexible programming structure.
- Data structuring on the basis of a tree.
- Automatic rollback option.

4. Java

It's a popular, easy-to-learn and quite versatile programming language, on the basis of which you can create apps of varying complexity for most operating systems.

Java virtual machine technology allows you to create one version of an app that will work on all Java-supported platforms. Its strong points are transparency, debugging convenience and portability.

Key features:

- Easy debugging.
- Simple to learn.
- Platform independence.
- Rich library.
- Scalability.

5. C++

One of the most important advantages of this programming language is its speed. This is important for developers, since the work of AI systems is accompanied by a large number of calculations, so speed can play a decisive role here. It is a good choice for projects based on machine learning and building neural networks.

Key features:

- Good speed and performance.
- A combination of high-level and low-level tools.
- Scalability.
- Plenty of available libraries and tools.

Unit 12

Text A

扫码听课文

The 10 Artificial Intelligence Trends

1. AI will increasingly be monitoring and refining business processes

While the first robots in the workplace were mainly involved with automating manual tasks such as manufacturing and production lines, today's software-based robots will take on the repetitive but necessary work that we carry out on computers. Filling in forms, generating reports and diagrams and producing documentation and instructions are all tasks that can be automated by machines that watch what we do and learn to do it for us in a quicker and more streamlined manner. This automation(known as robotic process automation)will free us from the drudgery of time-consuming but essential administrative work, leaving us to spend more time on complex, strategic, creative and interpersonal tasks.

2. More and more personalization will take place in real-time

This trend is driven by the success of internet giants like Amazon, Alibaba and Google, and their ability to deliver personalized experiences and recommendations. AI allows providers of goods and services to quickly and accurately project a 360-degree view of customers in real-time. As they interact through online portals and mobile apps, they quickly learn how their predictions can fit our wants and needs with ever-increasing accuracy. Just as pizza delivery companies like Dominos will learn when we are most likely to want pizza, and make sure the "Order Now" button is in front of us at the right time, every other industry will roll out solutions aimed at offering personalized customer experiences at scale.

3. AI will become increasingly useful as data becomes more accurate and available

The quality of information available is often a barrier to businesses and organizations wanting to move towards AI-driven automated decision-making. But as technology and methods of simulating real-world processes and mechanisms in the digital domain have improved over recent years, accurate data has become increasingly available. Simulations have advanced to the stage where car manufacturers and others working on the development of autonomous vehicles can gain thousands of hours of driving data without vehicles even leaving the lab, leading to huge reductions in cost as well as increases in the quality of data that can be gathered. Why risk the expense and danger of testing AI systems in the real world when computers are now powerful enough, and trained on accurate-enough data, to simulate it all in the digital world? This is because the accuracy and availability of real-world simulations will increase, which in turn will lead to more powerful and accurate AI.

4. More devices will run AI-powered technology

As the hardware and expertise needed to deploy AI become cheaper and more available, we will start to see it used in an increasing number of tools, gadgets and devices. We're already used to running apps that give us AI-powered predictions on our computers, phones and watches. As the cost of hardware and software continues to fall, AI tools will increasingly be embedded into our vehicles, household appliances and workplace tools. Augmented by technology such as virtual and augmented reality displays, and paradigms like the cloud and Internet of Things, we will see more and more devices of every shape and size starting to think and learn for themselves.

5. Human and AI cooperation will increase

More and more of us will get used to the idea of working alongside AI-powered tools and bots in day-to-day working lives. Increasingly, tools will be built that allow us to make the most of our human skills(those which AI can't quite manage yet)such as imagination, design, strategy and communication skills, while augmenting them with super-fast analytics abilities fed by vast datasets that are updated in real-time.

For many of us, this will mean learning new skills, or at least new ways to use our skills alongside these new robotic and software-based tools. The IDC predicts that by 2025, 75% of organizations will be investing in employee retraining in order to fill skill gaps caused by the need to adopt AI. This trend will become increasingly apparent throughout 2023, to the point where if your employer isn't investing in AI tools and training, it might be worth considering how well placed they are to grow over the coming years.

6. AI will be increasingly at the "edge"

Much of the AI we're used to interacting with now in our day-to-day lives takes place "in the cloud" — when we search on Google or flick through recommendations on Netflix, the complex, data-driven algorithms run on high-powered processors inside remote data centers, with the devices in our hands or on our desktops simply acting as conduits for information to pass through.

However, as these algorithms become more efficient and capable of running on low-power devices, AI is taking place at the "edge, " close to the point where data is gathered and used. This paradigm will become more popular in 2023 and beyond, making AI-powered insights a reality outside of the times and places where super-fast fiber optic and mobile networks are available. Custom processors designed to carry out real-time analytics on-the-fly will increasingly become part of the technology we interact with day-to-day, and increasingly we will be able to do this even if we have patchy or non-existent internet connections.

7. AI will be increasingly used to create films, music and games

We have seen AI-generated music, poetry or storytelling, however the effect is not ideal. It is generally agreed that even the most sophisticated machines still have some way to go until their output will be very enjoyable to us. However, the influence of AI on entertainment media is likely to increase. The use of AI in creating brand new visual effects is becoming increasingly common.

In videogames, AI will continue to be used to create challenging, human-like opponents for players to compete against, as well as to dynamically adjust gameplay and difficulty so that games can continue to offer a compelling challenge for gamers of all skill levels. And while completely AI-generated music may not be everyone's cup of tea, where AI does excel is in creating dynamic soundscapes — think of smart playlists on services like Spotify or Google Music that match tunes and tempo to the mood and pace of our everyday lives.

8. AI will become ever more present in cybersecurity

As hacking, phishing and social engineering attacks become ever-more sophisticated, and they are powered by AI and advanced prediction algorithms, smart technology will play an increasingly important role in protecting us from these attempted intrusions into our lives. AI can be used to spot signs that are likely to be indicators of nefarious activity, and raise alarms before defenses can be breached and sensitive data compromised.

The rollout of 5G and other super-fast wireless communications technology will bring huge opportunities for businesses to provide services in new and innovative ways, but they will also potentially open us up to more sophisticated cyber-attacks. Spending on cyber-security will continue to increase, and those with relevant skills will be highly sought-after.

9. More of us will interact with AI, maybe without even knowing it

Let's face it, despite the huge investment in recent years in natural-language powered chatbots in customer service, most of us can recognize whether we're dealing with a robot or a human. However, as the datasets used to train natural language processing algorithms continue to grow, the line between humans and machines will become harder and harder to distinguish. With the advent of deep learning and semi-supervised models of machine learning such as reinforcement learning, the algorithms that attempt to match our speech patterns and infer meaning from human language will become more and more able to fool us into thinking there is a human on the other end of the conversation. And while many of us may think we would rather deal with a human when looking for information or assistance, if robots fill their promise of becoming more efficient and accurate at interpreting our questions, that could change. Given the ongoing investment and maturation of the technology powering customer service bots and portals, 2023 could be the first time many of us interact with a robot without even realizing it.

10. AI will recognize us, even if we don't recognize it

Perhaps even more unsettlingly, the rollout of facial recognition technology is only likely to intensify as we move into the next decade. Corporations and governments are increasingly investing in these methods of telling who we are and interpreting our activity and behavior. But the question of whether people will ultimately begin to accept this intrusion into their lives in return for the increased security and convenience it will bring is likely to be a hotly debated topic.

✎ New Words

monitor	['mɒnɪtə]	v.监视；控制；监测
workplace	['wɜ:kpleɪs]	n.工作场所，车间；工厂
form	[fɔ:m]	n.表格；类型；形态，外形

report	[rɪ'pɔːt]	n.报告
diagram	['daɪəgræm]	n.图表，示意图
		v.用图表表示
documentation	[ˌdɒkjumen'teɪʃn]	n.文档
drudgery	['drʌdʒəri]	n.单调沉闷的工作
interpersonal	[ˌɪntə'pɜːsənl]	adj.人与人之间的；人际的
portal	['pɔːtl]	n.门户，入口
barrier	['bæriə]	n.障碍，屏障
hardware	['hɑːdweə]	n.硬件
gadget	['gædʒɪt]	n.小装置；小配件
software	['sɒftweə]	n.软件
paradigm	['pærədaɪm]	n.范式
update	[ˌʌp'deɪt]	vt.更新，升级
retrain	[ˌriː'treɪn]	vt.重新训练，再教育
gap	[gæp]	n.缺口；差距
processor	['prəʊsesə]	n.处理器，数据处理机
sophisticated	[sə'fɪstɪkeɪtɪd]	adj.先进的；复杂的；老练的；见多识广的
enjoyable	[ɪn'dʒɔɪəbl]	adj.有乐趣的；令人愉快的
videogame	['vɪdiəʊgeɪm]	n.电子游戏
opponent	[ə'pəʊnənt]	n.对手；敌手
gameplay	['geɪmpleɪ]	n.（计算机游戏的）游戏情节设计，玩法
soundscape	['saʊndskeɪp]	n.音景
tempo	['tempəʊ]	n.节奏，拍子
mood	[muːd]	n.情绪
intrusion	[ɪn'truːʒn]	n.入侵，打扰
spot	[spɒt]	v.注意到
sign	[saɪn]	n.迹象；符号
indicator	['ɪndɪkeɪtə]	n.指示器；指标
nefarious	[nɪ'feəriəs]	adj.极坏的，恶毒的
alarm	[ə'lɑːm]	v.报警
		n.警报（器）
defense	[dɪ'fens]	n.防御
breach	[briːtʃ]	v.突破，攻破
wireless	[ˌwaɪələs]	adj.无线的
communication	[kəˌmjuːnɪ'keɪʃn]	n.通信
cyber-attack	[ˌsaɪbə ə'tæk]	n.网络攻击
cyber-security	[ˌsaɪbə sɪ'kjuərəti]	n.网络安全
sought-after	['sɔːtɑːftə]	adj.很吃香的，广受欢迎的
conversation	[ˌkɒnvə'seɪʃn]	n.交谈，谈话
recognize	['rekəgnaɪz]	vt.认出；识别
unsettlingly	[ʌn'setlɪŋli]	adv.令人不安地

✎ Phrases

business process	业务流程，业务过程
production line	生产线，流水线
software-based robot	基于软件的机器人
fill in	填写；填补
personalized experience	个性化的体验
roll out	推出；铺开
household appliance	家用电器
augmented reality	增强现实
flick through	浏览
data-driven algorithm	由数据驱动的算法
high-powered processor	高性能处理器
low-power device	低功耗设备
smart playlist	智能播放列表
nefarious activity	恶意活动
sensitive data	敏感数据
wireless communications technology	无线通信技术
semi-supervised model	半监督模型
debated topic	争论的话题

Reference Translation

人工智能的十大趋势

1. 人工智能将越来越多地监控和完善业务流程

虽然工作场所的第一批机器人主要涉及自动化制造和生产线等手动任务，但如今基于软件的机器人将承担我们在计算机上执行的重复但必要的工作。填写表格、生成报告和图表以及生成文档和说明都是可以由机器自动完成的任务，这些机器会观察我们所做的事情并学会以更快、更流畅的方式为我们做这件事。这种自动化（被称为机器人过程自动化）将把我们从耗时但必不可少的行政工作中解放出来，让我们可以把更多的时间花在复杂、战略性、创造性和人际交往的任务上。

2. 越来越多的个性化将实时发生

这一趋势的推动因素是亚马逊、阿里巴巴和谷歌等互联网巨头的成功，以及它们提供个性化体验和推荐的能力。人工智能使商品和服务提供商能够快速准确地实时预测客户的360°视图。当他们通过在线门户和移动应用程序进行交互时，他们很快就会了解他们的预测如何能够以越来越高的准确性来满足我们的需求。就像 Dominos 之类的比萨外卖公司会了解我们什么时候最有可能想要比萨，并确保"立即订购"按钮在正确的时间出现在我们面前一样，其他所有行业都将大规模地推出旨在提供个性化客户体验的解决方案。

3. 随着数据变得更加准确和可用，人工智能将变得越来越有用

当企业和组织希望转向人工智能驱动的自动化决策时，可用信息的质量通常是一个障碍。但是，随着近年来在数字领域模拟现实世界过程和机制的技术与方法不断改进，准确的数据变得越来越容易获得。模拟已经发展到这样一个阶段：汽车制造商和其他致力于开发自动驾驶汽车的人可以在车辆甚至不离开实验室的情况下就获得数千小时的驾驶数据，从而大大降低了成本，提高了可收集数据的质量。既然计算机现在已经足够强大，并且经过足够准确的数据训练，可以在数字世界中进行模拟，为什么还要冒着花费资金和遇到危险的风险而在现实世界中测试人工智能系统呢？这是因为现实世界模拟的准确性和可用性将会提高，这反过来又会导致更强大和更准确的人工智能。

4. 更多设备将运行人工智能技术

随着部署人工智能所需的硬件和专业知识变得更廉价、更容易获得，我们将看到越来越多的工具、小装置和设备使用人工智能。我们已经习惯于在计算机、手机和智能手表上运行可以为我们提供基于人工智能预测的应用程序。随着硬件和软件成本的不断下降，人工智能工具将越来越多地嵌入到我们的车辆、家用电器和工作场所工具中。随着虚拟现实和增强现实显示等技术以及云和物联网等范式的发展，我们将看到越来越多的各种形状和大小的设备开始自行思考与学习。

5. 人与人工智能的合作将增加

越来越多的人将习惯于在日常工作生活中与人工智能驱动的工具和机器人一起工作。越来越多的工具将使我们能够充分利用我们的人类技能（那些人工智能还不能完全掌握的技能），如想象力、设计、战略和沟通技巧，同时通过大量实时更新的数据集提供的超快速分析能力来增强这些技能。

对我们中的许多人来说，这将意味着需要学习新的技能，或者至少学习新的方法，与这些新的机器人和基于软件的工具一起使用我们的技能。IDC 预测，到 2025 年，75%的组织将投资于员工再培训，以填补因需要采用人工智能而造成的技能人才缺口。这种趋势将在 2023 年及以后变得越来越明显，如果你的雇主不投资人工智能工具和培训，那么可能需要考虑他们在未来几年的发展情况。

6. 人工智能将越来越处于"边缘"

我们现在在日常生活中习惯与之交互的大部分人工智能都发生在"云端"——当我们在谷歌上搜索或浏览 Netflix 上的推荐时，复杂的数据驱动算法在远程数据中心内的高性能处理器上运行，我们手中或桌面上的设备只是充当信息传递的管道。

然而，随着这些算法变得越来越高效并且能够在低功耗设备上运行，人工智能正在靠近收集和使用数据的"边缘"发生。这种范式将在 2023 年及以后变得更加流行，使人工智能驱动的洞察力在超高速光纤和移动网络可用的时代和地点之外成为现实。旨在即时执行实时分析的定制处理器将越来越多地成为我们日常交互技术的一部分，即使互联网连接状况不好或没有联网，我们也能够做到这一点。

7. 人工智能将越来越多地用于制作电影、音乐和游戏

我们见过人工智能生成的音乐、诗歌或故事，但效果并不理想。人们普遍认为，即使最先进的机器也还有一段路要走，直到它们的输出对我们来说是非常令人愉快的。然而，人工智能对娱乐媒体的影响可能会增加。用人工智能创造全新的视觉效果越来越普遍。

在电子游戏中，人工智能将继续用于创造具有挑战性的、像人一样的对手以供玩家与之进行竞赛，并动态调整游戏玩法和难度，以便游戏能够继续为所有技能水平的游戏玩家提供引人入胜的挑战。虽然完全由人工智能生成的音乐可能不是每个人都喜欢的，但人工智能真正擅长的是创造动态音景——想想 Spotify 或 Google Music 等服务上的智能播放列表，它们可以根据我们日常生活中的情绪和节奏来匹配音乐的曲调与节奏。

8. 人工智能将越来越多地出现在网络安全领域

随着黑客、网络钓鱼和社会工程攻击变得越来越复杂，并且它们由人工智能和高级预测算法提供支持，智能技术将在保护我们的日常生活免受这些有意侵害方面发挥越来越重要的作用。人工智能可用于发现可能是恶意活动迹象的标志，并在防御系统被破坏和敏感数据泄露之前发出警报。

5G和其他超高速无线通信技术的推出将为企业以全新的创新方式提供服务带来巨大机遇，但它们也可能使我们面临更复杂的网络攻击。网络安全支出将继续增加，具有相关技能的人将受到高度追捧。

9. 我们中的更多人将与人工智能互动，甚至可能自己都不知道

让我们面对现实，尽管近年来在客户服务中对自然语言支持的聊天机器人进行了巨额投资，但我们中的大多数人都能识别出我们是在与机器人打交道，还是与人类打交道。然而，随着用于训练自然语言处理算法的数据集的不断增长，人与机器之间的界限将变得越来越难以区分。随着深度学习和机器学习半监督模型（如强化学习）的出现，试图匹配我们的语音模式并从人类语言中推断含义的算法将越来越能够欺骗我们，使我们认为对话的另一端是人类。虽然我们中的许多人可能认为在寻找信息或帮助时更愿意与人类打交道，但如果机器人履行其诺言，在解释我们的问题时变得更加高效和准确，那情况就可能会发生变化。鉴于支持客户服务机器人和门户的技术的成熟及对其进行的持续投资，在 2023 年，我们中的许多人可能会第一次在没有意识到的情况下与机器人互动。

10. 人工智能会认出我们，即使我们不认得它

也许更令人不安的是，随着我们进入下一个十年，面部识别技术的推广可能只会愈演愈烈。公司和政府会越来越多地投资于这些能够识别我们并解释我们的活动和行为的方法。但是，人们最终是否会开始接受这种对他们生活的入侵，以换取它所带来的安全性和便利性的提高，这个问题很可能是一个激烈争论的话题。

Text B

扫码听课文

The Role of Artificial Intelligence in Cloud Computing

1. Introduction

Cloud computing services have morphed from platforms such as Google App Engine and Azure to infrastructure which involves the provision of machines for computing and storage. In addition to this, cloud providers also offer data platform services which span the different available databases. This chain of development points in the direction of the growth of artificial intelligence and cloud computing.

2. The existing types of cloud application development services

(1) Infrastructure as a Service (IaaS)

This is the cloud app development service which is most employed by users. It allows you to pay based on the usage of the services provided, a truly flexible plan. The services provided include renting storage, networks, operating systems, servers and Virtual Machines (VM).

(2) Platform as a Service (PaaS)

This service is designed to make web creation and mobile app design easier by having an inbuilt infrastructure of servers, networks, databases and storage that eliminates the need to constantly update them or manage them.

(3) Software as a Service (SaaS)

With this, the cloud provider and not the user is tasked with management and maintenance and all the user has to do to gain access is to be connected to the application over the internet with a web browser on his phone, tablet or PC. The SaaS is available over the internet on demand or on a subscription basis.

3. Types of cloud deployment

(1) Public cloud

For public clouds like the Microsoft Azure, the cloud provider owns and manages all hardware, software and other supporting infrastructures and is responsible for delivering computing resources (servers, storage) over the internet. As a user, you gain access to these services and manage your account through the web browser.

(2) Private cloud

Just as the name implies, a private cloud's services and infrastructure are maintained on a private network either by the providing company or a hired third-party service provider. It is used by a single organization and is sometimes located in the company's on-site data center itself.

(3) Hybrid cloud

This is a fusion of both the public and the private cloud services. It is made available by the integration of the personalized data and applications shared by both platforms. Clients looking for more flexible cloud app development solutions and a wide range of deployment options are advised to embrace this technology.

4. The results from the Merger between AI and cloud computing

(1) AI infrastructure for cloud computing

We can generate Machine Learning (ML) models when a large set of data is applied to certain algorithms, and it becomes important to leverage the cloud for this. The models are able to learn from the different patterns which are gleaned from the available data.

As we provide more data for this model, the prediction gets better and the accuracy is improved. For instance, for ML models which identify tumors, thousands of radiology reports are used to train the system. This pattern can be used by any industry since it can be customized based on the project needs. The data is the required input and this comes in different forms, such as raw data, unstructured data, etc.

Because of the advanced computation techniques which require a combination of CPUs and

GPUs, cloud providers now provide virtual machines with incredibly powerful GPUs. Also, machine learning tasks are now being automated using services which include batch processing, serverless computing and container orchestration. IaaS also helps in handling predictive analytics.

(2) AI services for cloud computing

Even without creating a unique ML model, it is possible to enjoy services provided by the AI systems. For instance, text analytics, speech, vision and machine language translation are accessible to developers. They can simply integrate this into their development projects.

Although these services are generic and are not tailored to specific uses, cloud computing vendors are taking steps to ensure that this is constantly improved. Cognitive computing is a model which allows users to provide their personalized data which can be trained to deliver well-defined services. This way, the problem of finding the appropriate algorithm or the correct training model is eliminated.

5. Benefits of blending AI with cloud computing

(1) It creates more efficient IT environments

An undeniable benefit of combining AI and cloud computing is gaining the ability to redesign your entire IT infrastructure. With competition becoming increasingly cut-throat, businesses feel the need to constantly set higher benchmarks, which necessitates innovation for long-term sustainability. AI-optimized applications and infrastructure are, therefore, going to continue to be in high demand.

IT service providers have taken note and have started introducing IT platforms equipped with storage and interconnected resources, which help automate and accelerate IT workloads. AI-powered cloud platforms and infrastructure-optimization tools can help in preparing the business IT infrastructure for increased demand.

(2) It enables unhindered data access

AI in cloud computing can provide users with seamless data access. AI uses data to get things done, which makes it well-suited to cloud environments as they can hold large amounts of data. This means AI technology in the cloud prevents issues related to delay in accessibility and poor performance.

The cloud can also harness AI for better decision-making. It can learn from all the data collected and arrive at more accurate estimates or solve potential issues before they even arise. Further, AI can facilitate unhindered data transfer between on-site and cloud IT environments.

The cloud ecosystem, especially hybrid cloud solutions, need unrestricted data movement, connectivity and accessibility. AI can help organizations access and manage their data more effectively in such environments. Organizations can scale according to industry standards simply by optimizing their resources.

(3) It improves data management

The inclusion of AI in the cloud can lead to a more effective synthesis of data systems for identifying valuable information. This information can then be applied practically in business operations.

With the volume of data increasing with time, it is becoming clear that organizations need to be equipped with a responsive cloud environment.

AI can enable organizations to control large quantities of data, which can then be evaluated to make sense of it. This way, it improves the responsiveness of the business's cloud ecosystem, while also developing its own capabilities to further strengthen organizational productivity and performance.

(4) It provides analytics-backed insights

Improved analytics is another outcome of the combination of AI and cloud computing. Organizations stand to get a precise analysis of business-critical data to generate more meaningful and valuable information. This will eliminate the need to hire expert analysts. AI can, in fact, achieve far better outcomes at lowered prices.

In addition, making sense of reams of numbers and hypotheses can require businesses to engage multiple teams that produce similar data repeatedly to identify new insights. AI can simplify these types of tasks, while also deriving accurate forecasts from colossal datasets. It can also identify patterns in data efficiently and provide recommendations backed by thorough analysis.

(5) It lowers costs

The merging of AI and cloud computing also has economic benefits in that it leads to reduced costs that may otherwise have to be incurred by businesses. This is mainly because expenses related to setting up and maintaining on-site data centers are eliminated.

AI can also bring down research and development-related costs. Organizations with access to the cloud can employ AI and gain meaningful insights without incurring additional outlays.

Since cloud solutions are scalable, they work well with the lowered cost-related concerns of AI-powered growth and development.

(6) It automates cloud security

AI has the capability of processing large amounts of information stored in the cloud and detect discrepancies instantaneously. This technology can automatically send out a system alert or respond in other ways in such cases. This, in turn, can instantly help block unauthorized access to and activity on cloud platforms.

Moreover, AI can identify atypical happenings across the network and block them right away, thereby thwarting attempts of potentially unsafe codes from entering the system. It can also help in reviewing data across locations, which can enable businesses to facilitate a hands-on response towards fortifying the security of business systems.

(7) It enhances decision-making

As mentioned, organizations that harness the power of the AI are able to identify patterns and trends in vast datasets. AI does this successfully by referring historical data and comparing it with the current information, thereby equipping you with well-informed, data-backed intelligence.

Output created by AI processes comes with higher accuracy as it does not involve human intervention, thereby removing man-made errors in data analysis. AI technology enables quicker data analysis, which means businesses can resolve customer queries promptly and more efficiently.

6. Conclusion

As you can see, AI and the cloud make the perfect marriage. They complement each other in many ways. In fact, AI is on its way to completely revolutionizing cloud computing technology. It is becoming increasingly evident that AI augments cloud services and creates new avenues for development.

✍ New Words

morph	[mɔ:f]	vt.改变
provision	[prə'vɪʒn]	v.为……提供所需物品
chain	[tʃeɪn]	n.链条
span	[spæn]	v.横跨，跨越
plan	[plæn]	n.计划；规划
network	['netwɜ:k]	n.网络
server	['sɜ:və]	n.服务器
subscription	[səb'skrɪpʃn]	n.订阅
browser	['braʊzə]	n.浏览器
on-site	['ɒn saɪt]	adj.现场的
application	[ˌæplɪ'keɪʃn]	n.应用
client	['klaɪənt]	n.客户
embrace	[ɪm'breɪs]	v.拥抱；欣然接受
merger	['mɜ:dʒə]	n.融合，合并
leverage	['li:vərɪdʒ]	n.杠杆作用
		v.利用
orchestration	[ˌɔːkɪ'streɪʃn]	n.编排；管弦乐编曲
container	[kən'teɪnə]	n.容器
generic	[dʒə'nerɪk]	adj.通用的；一般的
deliver	[dɪ'lɪvə]	v.交付；发表；递送
undeniable	[ˌʌndɪ'naɪəbl]	adj.不可否认的，无可争辩的
cut-throat	['kʌt θrəʊt]	adj.竞争激烈的；残酷无情的
benchmark	['bentʃmɑ:k]	n.基准；参照
sustainability	[sə,steɪnə'bɪləti]	n.持续性
workload	['wɜ:kləʊd]	n.工作量，工作负担
unhindered	[ʌn'hɪndəd]	adj.不受妨碍的，不受阻碍的
seamless	['si:mləs]	adj.无缝的
delay	[dɪ'leɪ]	v.延迟，推迟，耽误
unrestricted	[ˌʌnrɪ'strɪktɪd]	adj.不受限制的；无限制的；自由的
synthesis	['sɪnθəsɪs]	n.综合
responsiveness	[rɪ'spɒnsɪvnəs]	n.响应性
strengthen	['streŋθn]	v.加强；巩固；支持；壮大
outcome	['aʊtkʌm]	n.结果
engage	[ɪn'geɪdʒ]	v.（使）参与；吸引
colossal	[kə'lɒsl]	adj.巨大的，庞大的
outlay	['aʊtleɪ]	n.花费；费用
scalable	['skeɪləbl]	adj.可扩展的，可升级的
discrepancy	[dɪ'skrepənsi]	n.差异，不一致
instantaneously	[ˌɪnstən'teɪniəsli]	adv.即刻地
unauthorized	[ʌn'ɔ:θəraɪzd]	adj.未经授权的；未经许可的

unsafe	[ʌn'seɪf]	*adj.*不安全的，危险的

✎ Phrases

operating system	操作系统
cloud provider	云技术提供商
public cloud	公共云
computing resource	计算资源
private cloud	私有云，专用云
private network	私有网络，专用网络
third-party service provider	第三方服务提供商
data center	数据中心
hybrid cloud	混合云
personalized data	个性化数据
raw data	原始数据
batch processing	批处理
serverless computing	无服务器计算
container orchestration	容器编排
cognitive computing	认知计算
industry standard	行业标准
business-critical data	关键业务数据
economic benefit	经济效益，经济利益
unauthorized access	未授权的访问
historical data	历史数据

✎ Abbreviations

IaaS (Infrastructure as a Service)	基础设施即服务
PaaS (Platform as a Service)	平台即服务
SaaS (Software as a Service)	软件即服务
CPU (Central Processing Unit)	中央处理器

Reference Translation

人工智能在云计算中的作用

1. 简介

    云计算服务已经从 Google App Engine 和 Azure 等平台转变为涉及提供计算与存储机器的基础设施。除此之外，云提供商还提供跨越不同可用数据库的数据平台服务。这一发展链指向了人工智能和云计算的增长方向。

2. 现有的云应用开发服务类型

    （1）基础设施即服务（IaaS）

    这是用户最常使用的云应用开发服务。它允许你根据所提供服务的使用情况付费，这是一个真正灵活的计划。其提供的服务包括租用存储、网络、操作系统、服务器和虚拟机（VM）。

（2）平台即服务（PaaS）

该服务旨在通过内置服务器、网络、数据库和存储基础设施，消除不断更新或管理它们的需要，从而使创建网络和移动应用程序设计变得更容易。

（3）软件即服务（SaaS）

使用此项，云提供商而不是用户负责管理和维护，用户获得访问权只需要通过手机、平板计算机或 PC 上的网络浏览器连接到互联网上的应用程序。软件即服务可通过互联网按需或订阅提供。

3. 云部署的类型

（1）公共云

对于像 Microsoft Azure 这样的公共云，云提供商拥有并管理所有硬件、软件和其他支持基础设施，并负责通过互联网提供计算资源，如服务器、存储。作为用户，你可以通过网络浏览器访问这些服务并管理你的账户。

（2）私有云

顾名思义，私有云的服务和基础设施由提供公司或聘请的第三方服务提供商在私有网络上维护。它由单个组织使用，有时就位于公司的现场数据中心。

（3）混合云

这是公共和私有云服务的融合。它是通过集成两个平台共享的个性化数据和应用程序来提供的。建议寻求更灵活的云应用程序开发解决方案和广泛部署选项的客户采用此技术。

4. 人工智能与云计算融合的结果

（1）用于云计算的人工智能基础设施

当大量数据应用于某些算法时，我们可以生成机器学习（ML）模型，因此利用云变得很重要。这些模型能够从可用数据收集的不同模式中学习。

随着我们为这个模型提供更多的数据，预测会变得更好，准确性也会提高。例如，对于识别肿瘤的机器学习模型，数以千计的放射学报告被用来训练系统。这种模式可用于任何行业，因为它可以根据项目需求进行定制。数据是必需的输入，它有不同的形式，如原始数据、非结构化数据等。

由于先进的计算技术需要 CPU 和 GPU 的结合，云提供商现在为虚拟机提供了功能强大的 GPU。此外，机器学习任务现在正在使用包括批处理、无服务器计算和容器编排在内的服务来实现自动化。基础设施即服务还有助于处理预测分析。

（2）用于云计算的人工智能服务

即使不创建独特的机器学习模型，也可以享受人工智能系统提供的服务。例如，开发人员可以访问文本分析、语音、视觉和机器语言翻译。他们可以简单地将其集成到他们的开发项目中。

尽管这些服务是通用的并且不是针对特定用途量身定制的，但云计算供应商正在采取措施确保不断改进。认知计算是一种模型，它允许用户提供他们的个性化数据，可以训练这些数据来提供定义明确的服务。这样，就消除了寻找合适算法或正确训练模型的问题。

5. 人工智能与云计算相结合的好处

（1）创建更高效的 IT 环境

人工智能与云计算结合的一个不可否认的好处是获得了重新设计整个 IT 基础设施的能力。随着竞争变得越来越激烈，企业感到有必要不断设定更高的基准，这就需要创新来实现长期可持续性。因此，对人工智能优化的应用程序和基础设施的需求将继续增大。

IT 服务提供商已经注意到并开始引入配备有存储和互联资源的 IT 平台，这有助于自动化和加速 IT 工作负载。人工智能驱动的云平台和基础设施优化工具可以帮助企业 IT 基础设施为不断增长的需求做好准备。

（2）实现无阻碍的数据访问

云计算中的人工智能可以为用户提供无缝的数据访问。人工智能使用数据来完成任务，这使得它非常适合云环境，因为它们可以保存大量数据。这意味着云中的人工智能技术可以防止与可访问性延迟和性能不佳相关的问题。

云还可以利用人工智能做出更好的决策。它可以从收集到的所有数据中学习，并在潜在问题出现之前做出更准确的估计或解决它们。此外，人工智能可以确保现场和云 IT 环境之间的数据传输不受阻碍。

云生态系统，尤其是混合云解决方案，需要不受限制的数据移动、连接和可访问性。人工智能可以帮助组织在这种环境中更有效地访问和管理它们的数据。组织只需优化其资源，即可根据行业标准进行扩展。

（3）改进数据管理

在云中包含人工智能可以更有效地综合数据系统，从而识别有价值的信息。然后，这些信息可以实际应用于业务运营。

因为数据量随时间的推移而增加，很明显，组织需要配备响应式云环境。

人工智能使组织能够控制大量数据，然后可以对其进行评估以了解其意义。通过这种方式，它提高了企业云生态系统的响应能力，同时还开发了自己的能力，以进一步增强组织的生产力和绩效。

（4）它提供分析支持的洞察力

改进的分析是人工智能和云计算结合的另一个结果。组织需要对关键业务数据进行精确分析，以生成更有意义和更有价值的信息。这将不需要聘请专家分析师。事实上，人工智能可以以更低的价格获得更好的结果。

此外，要理解大量的数字和假设，企业需要多个团队反复提供类似的数据，以确定新的见解。人工智能可以简化这些类型的任务，同时还可以从庞大的数据集中得出准确的预测。它还可以有效地识别数据中的模式，然后进行全面分析并提出建议。

（5）降低成本

人工智能和云计算的融合还具有经济效益，因为它可以降低企业可能不得不承担的成本。这主要是因为消除了与建立和维护现场数据中心相关的费用。

人工智能还可以降低与研发相关的成本。能够访问云的组织可以使用人工智能并获得有意义的见解，而不会产生额外的支出。

由于云解决方案是可扩展的，因此它们可以很好地解决人工智能驱动的增长和开发中与成本相关的问题。

（6）使云安全自动化

人工智能能够处理存储在云中的大量信息并即时检测差异。在这种情况下，该技术可以自动发出系统警报或以其他方式响应。反过来，这可以立即帮助阻止对云平台的未经授权的访问和活动。

此外，人工智能可以识别网络上的非典型事件并立即阻止它们，从而阻止潜在的不安全代码进入系统的企图。它还可以帮助跨位置审查数据，这可以使企业能够促进实际响应，以加强业务系统的安全性。

（7）增强决策能力

如前所述，利用人工智能力量的组织能够识别大量数据集中的模式和趋势。人工智能可能通过参考历史数据并将其与当前信息进行比较来成功地做到这一点，从而为人们提供消息灵通、有数据支持的情报。

人工智能过程创建的输出具有更高的准确性，因为它不涉及人工干预，从而消除了数据分析中的人为错误。人工智能技术可实现更快的数据分析，这意味着企业可以迅速、更有效地解决客户查询。

6. 结论

如你所见，人工智能和云是完美的结合。它们在许多方面相互补充。事实上，人工智能正在彻底改变云计算技术。越来越明显的是，人工智能增强了云服务并创造了新的发展途径。

Exercises

[Ex. 1] **Answer the following questions according to Text A.**

1. What is this automation known as? What will this automation do?

2. What does AI allow providers of goods and services to do?

3. What is often a barrier to businesses and organizations wanting to move towards AI-driven automated decision-making?

4. What are we already used to?

5. What does the IDC predict?

6. Where does much of the AI we're used to interacting with now in our day-to-day lives take place?

7. In videogames, what will AI do?

8. What will the rollout of 5G and other super-fast wireless communications technology do?

9. What will the algorithms that attempt to match our speech patterns and infer meaning from our own human language do with the advent of deep learning and semi-supervised models of machine learning such as reinforcement learning?

10. What is likely to be a hotly debated topic?

[Ex. 2] **Answer the following questions according to Text B.**

1. What have cloud computing services done?

2. What are the existing types of cloud application development services?

3. What is PaaS designed to do?

4. What are the types of cloud deployment?

5. Why do cloud providers now provide virtual machines with incredibly powerful GPU?

6. What is cognitive computing?

7. What is an undeniable benefit of combining AI and cloud computing?

8. What can the inclusion of AI in the cloud lead to?

9. Why does the merging of AI and cloud computing also have economic benefits?

10. What can AI do?

[Ex. 3] Translate the following terms or phrases from English into Chinese.

1. communication
2. cyber-security
3. intrusion
4. monitor
5. paradigm
6. augmented reality
7. semi-supervised model
8. personalized experience
9. software-based robot
10. computing resource

[Ex. 4] Translate the following terms or phrases from Chinese into English.

1. *n.*处理器，数据处理机
2. *vt.*认出；识别
3. *adj.*无线的
4. *n.*基准；参照
5. *n.*网络
6. 认知计算
7. 云技术提供商
8. 混合云
9. 操作系统
10. 原始数据

[Ex. 5] Translate the following passage into Chinese.

How Big Data and AI Will Work Together

Both big data and AI applications have a mutually beneficial relationship. The success of AI applications depends on the big data input. AI is now assisting organizations towards utilizing their data as a means of influencing organizational decision-making with previously thought-to-be unfeasible methodologies.

There are three key ways that AI can deliver better insights with big data.

1. AI is creating new and enhanced methods for analyzing data

Determining insight from data previously required much manual effort from an organization's staff. Historically, engineers had to use a SQL query or a list of SQL queries to analyze data. With AI, an array of new and enhanced methods to obtain data insights have become available. Thus, AI and machine learning are now creating new and more efficient methods for analyzing an immense quantity of data.

2. AI can be used to alleviate common data problems

The value of big data sets is intricately tied to data quality. Data that is of deficient quality is of little or no worth for the organizational decision-making process. Machine learning algorithms in AI applications can discover outlier and missing values, duplicate records and standardize data for big data analytics.

3. Analytics become more predictive and prescriptive

In the past, data analytics was primarily backward-looking with a post-analysis of what happened. Predictions and forecasts were essentially historical analyses. Big data decisions were therefore based on past and present data points with a linear ROI.

AI is now creating new opportunities for enhanced predictions and forecasts. An AI algorithm can be set up to make a decision or take an action based upon forward-looking insights. In essence, big data analytics can become more predictive and prescriptive.

附录

附录 A 单 词 表

单词	音标	意义	单元
abductive	[æbˈdʌktɪv]	adj.溯因的；诱导的	3A
ability	[əˈbɪləti]	n.能力；才能	1A
absorb	[əbˈzɔːb]	v.吸收；理解，掌握	6B
abundantly	[əˈbʌndəntli]	adv.丰富地；大量地	5B
abuse	[əˈbjuːs]	n.&v.滥用	5B
acceptable	[əkˈseptəbl]	adj.可接受的	3B
accommodate	[əˈkɒmədeɪt]	v.适应	10A
accumulate	[əˈkjuːmjəleɪt]	v.积累，积聚	8B
accuracy	[ˈækjərəsi]	n.精确（性），准确（性）	5A
accurately	[ˈækjərətli]	adv.准确地；精确地	9A
achieve	[əˈtʃiːv]	v.实现，达到	11B
activation	[ˌæktɪˈveɪʃn]	n.激活	8A
active	[ˈæktɪv]	adj.积极的；活跃的	5B
actuator	[ˈæktʃʊeɪtə]	n.执行器	10A
adapt	[əˈdæpt]	v.（使）适应/适合	9A
adjust	[əˈdʒʌst]	v.调整，调节；适应；校准	6A
adjustable	[əˈdʒʌstəbl]	adj.可调整的，可调节的	8A
advancement	[ədˈvɑːnsmənt]	n.发展，推动	1A
advise	[ədˈvaɪz]	v.劝告；建议	5A
advisor	[ədˈvaɪzə]	n.顾问	5A
affective	[əˈfektɪv]	adj.情感的	1A
affordable	[əˈfɔːdəbl]	adj.价格合理的	10B
agent	[ˈeɪdʒənt]	n.实体；代理	2A
aggregate	[ˈægrɪgət] [ˈægrɪgeɪt]	n.总数，合计 adj.总数的，总计的 v.总计；汇集	3B
aid	[eɪd]	n.&v.帮助，辅助	1A
alarm	[əˈlɑːm]	v.报警 n.警报（器）	12A
algorithm	[ˈælgərɪðəm]	n.算法	4A
ambiguity	[ˌæmbɪˈgjuːəti]	n.歧义；不明确，模棱两可	9B
ambiguous	[æmˈbɪgjuəs]	adj.模棱两可的；不明确的	6B

214

单　词	音　标	意　义	单　元
analogy	[əˈnælədʒi]	n.类推	1A
annotation	[ˌænəˈteɪʃn]	n.注释	2B
anxiety	[æŋˈzaɪəti]	n.忧虑，焦虑	5A
appeal	[əˈpiːl]	v.对……有吸引力　n.吸引力	11A
application	[ˌæplɪˈkeɪʃn]	n.应用	12B
apply	[əˈplaɪ]	v.应用，适用	1A
approximation	[əˌprɒksɪˈmeɪʃn]	n.近似法，近似值	4A
argument	[ˈɑːɡjumənt]	n.论据；讨论，辩论；争论	3A
aspect	[ˈæspekt]	n.方面；样子	4A
aspiration	[ˌæspəˈreɪʃn]	n.抱负，志向	11A
assess	[əˈses]	v.评估；估价；估算	9A
assign	[əˈsaɪn]	v.指派；给予	9A
association	[əˌsəʊsiˈeɪʃn]	n.关系；联系；因果关系	6A
assume	[əˈsjuːm]	v.假设，假定	5A
assumption	[əˈsʌmpʃn]	n.假定；承担；获得	3A
astound	[əˈstaʊnd]	vt.使震惊，使大吃一惊	4A
attest	[əˈtest]	v.证实，证明	8B
attribute	[əˈtrɪbjuːt]	n.属性，性质，特征	2A
auto-correct	[ˈɔːtəʊ kərekt]	v.自动校正	1B
automation	[ˌɔːtəˈmeɪʃn]	n.自动化	1B
autonomous	[ɔːˈtɒnəməs]	adj.自治的，自主的	10A
autonomy	[ɔːˈtɒnəmi]	n.自主（权）	10A
available	[əˈveɪləbl]	adj.可用的，可获得的；有空的	1A
awareness	[əˈweənəs]	n.意识；了解；觉察	2B
backend	[bækend]	n.后端	10B
background	[ˈbækɡraʊnd]	n.背景	9A
backtrack	[ˈbæktræk]	vi.回溯	4B
balance	[ˈbæləns]	n.均衡，平衡（能力）	1B
barrier	[ˈbæriə]	n.障碍，屏障	12A
basis	[ˈbeɪsɪs]	n.基础；基准，根据；方式	10B
bear	[beə]	v.带有，拥有	2B
behavioral	[bɪˈheɪvjərəl]	adj.行为的	6A
benchmark	[ˈbentʃmɑːk]	n.基准，参照	12B
beneficial	[ˌbenɪˈfɪʃl]	adj.有益的，有帮助的	1A
bias	[ˈbaɪəs]	n.偏好；偏见	6A
binary	[ˈbaɪnəri]	adj.二态的；二元的	7B
biopsy	[ˈbaɪɒpsi]	n.活组织检查，活体检视	9A

（续）

单 词	音 标	意 义	单 元
blind	[blaɪnd]	adj.盲目的；不理性的	4B
blockchain	['blɒkt'ʃeɪn]	n.区块链	10B
boolean	['bu:liən]	adj.布尔的	3B
boost	[bu:st]	vt.促进，提高；增加	1B
bootstrap	['bu:tstræp]	n.引导程序	2B
borrow	['bɒrəʊ]	v.借用，引用	2B
brake	[breɪk]	n.制动器	10A
breach	[bri:tʃ]	v.突破，攻破	12A
breadthwise	['bredθwaiz]	adv.横向地	4B
breakdown	['breɪkdaʊn]	n.故障	7B
browser	['braʊzə]	n.浏览器	12B
builder	['bɪldə]	n.构建器；开发器	5B
built-in	[,bɪlt'ɪn]	adj.嵌入的；内置的	11B
calculator	['kælkjuleɪtə]	n.计算器；计算者	1A
capability	[,keɪpə'bɪləti]	n.能力	1A
capable	['keɪpəbl]	adj.有能力的；胜任的	6B
capture	['kæptʃə]	v.捕捉，捕获	2B
categorisation	[,kætɪgəraɪ'zeɪʃn]	n.分类	5B
categorize	['kætəgəraɪz]	vt.把……归类，把……分门别类	2A
chain	[tʃeɪn]	n.链条	12B
characteristic	[,kærəktə'rɪstɪk]	n.特色；特点	5A
chat	[tʃæt]	v.&n.闲谈；聊天	1B
chatbot	[tʃætbɒt]	n.聊天机器人	6A
check	[tʃek]	v.检查；查看；核实	4A
class	[klɑ:s]	n.类	2B
classification	[,klæsɪfɪ'keɪʃn]	n.分类，类别	9A
classify	['klæsɪfaɪ]	v.将……分类	7A
client	['klaɪənt]	n.客户	12B
clue	[klu:]	n.线索；提示	9B
cluster	['klʌstə]	v.聚集 n.团，群，簇	6A
code	[kəʊd]	n.代码；编码 v.把……编码；编程序	5B
coder	['kəʊdə]	n.程序员，编程者；编码器	5B
codify	['kəʊdɪfaɪ]	v.编纂；整理；将（法规等）整理成典	5A
collaborate	[kə'læbəreɪt]	v.合作，协作	1B
collaborative	[kə'læbərətɪv]	adj.合作的，协作的	11A
collection	[kə'lekʃn]	n.集；群	5A
collectively	[kə'lektɪvli]	adv.全体地，共同地	10B

（续）

单　词	音　标	意　义	单　元
colossal	[kə'lɒsl]	adj.巨大的，庞大的	12B
combination	[ˌkɒmbɪ'neɪʃn]	n.结合（体）；联合（体）	4A
combine	[kəm'baɪn]	v.使结合，使合并	4A
command	[kə'mɑːnd]	n.&v.命令；控制	9B
communication	[kəˌmjuːnɪ'keɪʃn]	n.通信	12A
community	[kə'mjuːnəti]	n.社区，社团	11A
compatible	[kəm'pætəbl]	adj.兼容的	5B
complete	[kəm'pliːt]	adj.完全的，完整的	4A
completeness	[kəm'pliːtnəs]	n.完备性；完全性；完整性	4B
complex	['kɒmpleks]	adj.复杂的	5A
complexity	[kəm'pleksəti]	n.复杂度，复杂性	4B
complicate	['kɒmplɪkeɪt]	v.使复杂化	5B
component	[kəm'pəʊnənt]	n.部件，组件　adj.组成的，构成的	9A
comprehend	[ˌkɒmprɪ'hend]	v.理解，领悟	3B
comprehensive	[ˌkɒmprɪ'hensɪv]	adj.全面的；综合性的	8B
compromise	['kɒmprəmaɪz]	n.折中；妥协方案；达成协议　v.妥协，使陷入危险，损害；泄露	3B
concept	['kɒnsept]	n.概念；观念	4A
conceptualize	[kən'septʃuəlaɪz]	vi.概念化	7B
concern	[kən'sɜːn]	v.考虑；关心；影响；涉及	4B
concise	[kən'saɪs]	adj.简明的，简洁的	11A
concurrency	[kən'kʌrənsi]	n.并发（性）	11A
condition	[kən'dɪʃn]	n.状况；环境；条件	7A
conditionally	[kən'dɪʃənəli]	adv.有条件地	7A
conduct	[kən'dʌkt]	v.引导，组织，实施	10A
confident	['kɒnfɪdənt]	adj.坚信的；自信的	11A
confirmation	[ˌkɒnfə'meɪʃn]	n.证实；确认	3A
confluence	['kɒnfluəns]	n.（事物的）汇合；汇流	1A
confusing	[kən'fjuːzɪŋ]	adj.难以理解的	7B
connectivity	[kəˌnek'tɪvəti]	n.连通性	10B
consequence	['kɒnsɪkwəns]	n.结果；重要性	1A
consistency	[kən'sɪstənsi]	n.一致性，连贯性	11A
consistent	[kən'sɪstənt]	adj.一致的；连贯的	11A
constraint	[kən'streɪnt]	n.约束，限制	7B
construct	[kən'strʌkt]	v.构造；组成	2A
container	[kən'teɪnə]	n.容器	12B
contextual	[kən'tekstʃuəl]	adj.上下文的；与语境有关的	9B
continue	[kən'tɪnjuː]	vi.持续，连续	1A

（续）

单　词	音　标	意　义	单　元
continuous	[kən'tɪnjuəs]	adj.连续不断的	8B
continuously	[kən'tɪnjuəsli]	adv.连续不断地	6A
contradictory	[ˌkɒntrə'dɪktəri]	adj.对立的	3A
control	[kən'trəʊl]	v.控制	3B
convenient	[kən'viːniənt]	adj.实用的；方便的	5B
converge	[kən'vɜːdʒ]	v.汇集，聚集；收敛	7A
conversation	[ˌkɒnvə'seɪʃn]	n.交谈，谈话	12A
conversational	[ˌkɒnvə'seɪʃənl]	adj.会话的	11B
convert	[kən'vɜːt]	v.转换，转变；改造	3B
convey	[kən'veɪ]	v.表达；传送	2A
corrective	[kə'rektɪv]	adj.改正的；矫正的	10B
correctness	[kə'rektnəs]	n.正确性	2A
coverage	['kʌvərɪdʒ]	n.覆盖范围	2B
craft	[krɑːft]	v.精心制作　n.手艺，技巧	5B
crawl	[krɔːl]	v.爬行	11A
creative	[kri'eɪtɪv]	adj.创造性的，有创意的，创新的	5A
crisp	[krɪsp]	adj.清晰的；洁净的；挺括的	3B
crowdsource	[kraʊdsɔːs]	vt.众包	2B
cumulative	['kjuːmjələtɪv]	adj.积累的，累计的	4B
curation	[ˌkjuː'reɪʃn]	n.管理；综合处理	5B
customize	['kʌstəmaɪz]	vt.定制，定做；用户化	11B
cut-throat	['kʌt θrəʊt]	adj.竞争激烈的；残酷无情的	12B
cyber	['saɪbə]	adj.计算机（网络）的，信息技术的	6A
cyber-attack	[ˌsaɪbə ə'tæk]	n.网络攻击	12A
cyber-security	[ˌsaɪbə sɪ'kjuərəti]	n.网络安全	12A
dangerous	['deɪndʒərəs]	adj.危险的	5B
database	['deɪtəbeɪs]	n.数据库	2A
data-based	['deɪtəbeɪst]	adj.以数据为基础，基于数据（的）	5B
dataset	['deɪtəset]	n.数据集	3B
debug	[ˌdiː'bʌg]	vt.调试；排除故障	5B
debugger	[ˌdiː'bʌgə]	n.调试器，调试程序	5B
decision	[dɪ'sɪʒn]	n.决策，决定	1A
decision-making	[dɪ'sɪʒnmeɪkɪŋ]	n.决策	5A
declarative	[dɪ'klærətɪv]	adj.陈述的	2A
deductive	[dɪ'dʌktɪv]	adj.推论的，演绎的	3A
dedupe	[di'djuːp]	v.删除重复数值	6A
default	[di'fɔːlt]	v.违约	7B

（续）

单　　词	音　　标	意　　义	单　元
defense	[dɪˈfens]	n.防御	12A
define	[dɪˈfaɪn]	v.定义；阐明；限定	6A
definite	[ˈdefɪnət]	adj.不会改变的；明确的	2A
definition	[ˌdefɪˈnɪʃn]	n.定义	4A
defuzzification	[dɪfʌzɪfɪˈkeɪʃn]	n.逆模糊化，去模糊化	3B
degree	[dɪˈgriː]	n.级别；程度	3B
delay	[dɪˈleɪ]	v.延迟，推迟，耽误	12B
deliver	[dɪˈlɪvə]	v.交付；发表；递送	12B
demonstrate	[ˈdemənstreɪt]	v.证明，说明；演示	2A
deploy	[dɪˈplɔɪ]	v.部署	11B
deployment	[dɪˈplɔɪmənt]	n.部署	11A
derive	[dɪˈraɪv]	v.（使）起源于，来自；获得	5A
describe	[dɪˈskraɪb]	v.描述；把……称为	5A
description	[dɪˈskrɪpʃn]	n.描述，说明；种类，性质	6B
detect	[dɪˈtekt]	v.发现；查明；测出	11A
deterministic	[dɪˌtɜːmɪˈnɪstɪk]	adj.确定性的	5A
developer	[dɪˈveləpə]	n.（产品等的）开发者	5B
development	[dɪˈveləpmənt]	n.开发；研制	5A
device	[dɪˈvaɪs]	n.设备，装置	10A
diagnosis	[ˌdaɪəɡˈnəʊsɪs]	n.诊断；判断	5A
diagram	[ˈdaɪəɡræm]	n.图表，示意图　　v.用图表表示	12A
dialogue	[ˈdaɪəlɒɡ]	n.对话	6B
dictation	[dɪkˈteɪʃn]	n.口述；听写	9B
digest	[daɪˈdʒest]	v.消化；理解	9B
digitization	[ˌdɪdʒɪtaɪˈzeɪʃən]	n.数字化	6B
diminish	[dɪˈmɪnɪʃ]	v.（使）减小，减弱	8B
disambiguation	[ˌdɪsæmˌbɪgjʊˈeɪʃən]	n.消歧，消除模棱两可情况	9B
discrepancy	[dɪˈskrepənsi]	n.差异，不一致	12B
discriminative	[dɪsˈkrɪmɪnətɪv]	adj.有判别力的	7A
disease	[dɪˈziːz]	n.疾病	5A
display	[dɪˈspleɪ]	v.显示	5A
disruptive	[dɪsˈrʌptɪv]	adj.颠覆性的，破坏性的	10B
distillation	[dɪstɪˈleɪʃn]	n.蒸馏（过程）；蒸馏物	3B
distinct	[dɪˈstɪŋkt]	adj.不同的；清楚的，明显的	8A
distinguish	[dɪˈstɪŋgwɪʃ]	v.区分，使有别于；辨别出	8A
distort	[dɪˈstɔːt]	v.歪曲，曲解；（使）变形，失真	3B
distributed	[dɪsˈtrɪbjuːtɪd]	adj.分布式的	11A

（续）

单 词	音 标	意 义	单 元
documentation	[ˌdɒkjumen'teɪʃn]	n.文档	12A
dominate	['dɒmɪneɪt]	v.控制，支配	11A
drudgery	['drʌdʒəri]	n.单调沉闷的工作	12A
durable	['djʊərəbl]	adj.耐用的；持久的	3B
earn	[ɜːn]	v.赚得；获得	1B
ecosystem	['iːkəʊsɪstəm]	n.生态系统	10B
effector	[ɪ'fektə]	n.效应器	10A
efficiency	[ɪ'fɪʃnsi]	n.效率，效能	5A
elevator	['elɪveɪtə]	n.电梯	3B
embed	[ɪm'bed]	v.（使）嵌入，融入	10B
embrace	[ɪm'breɪs]	v.拥抱；欣然接受	12B
emerge	[i'mɜːdʒ]	v.出现，兴起	1A
emotion	[ɪ'məʊʃn]	n.情绪	5A
emphasis	['emfəsɪs]	n.重读	9B
empty	['empti]	adj.空的	4A
emulate	['emjuleɪt]	v.模拟；仿效，模仿	8A
enable	[ɪ'neɪbl]	v.使能够；使可行	5B
enabler	[ɪ'neɪblə]	n.推动者	10B
encounter	[ɪn'kaʊntə]	v.遭遇；偶遇	6A
encyclopedic	[ɪnˌsaɪklə'piːdɪk]	adj.百科全书的	2B
endless	['endləs]	adj.无穷尽的；无止境的	9B
enforcement	[ɪn'fɔːsmənt]	n.实施，执行	6B
engage	[ɪn'geɪdʒ]	v.（使）参与；吸引	12B
enjoyable	[ɪn'dʒɔɪəbl]	adj.有乐趣的；令人愉快的	12A
enormous	[ɪ'nɔːməs]	adj.巨大的；极大的	1A
enrich	[ɪn'rɪtʃ]	v.使充实；使丰富	2B
ensemble	[ɒn'sɒmbl]	n.集成；全体，整体	7B
ensure	[ɪn'ʃʊə]	v.确保；担保	5B
entertainment	[ˌentə'teɪnmənt]	n.娱乐节目，娱乐活动	7A
entity	['entəti]	n.实体	1A
epidemiology	[ˌepɪˌdiːmɪ'ɒlədʒi]	n.流行病学	7B
equivalent	[ɪ'kwɪvələnt]	adj.相等的，相同的　　n.等同物；对应物	2A
error-free	['erə friː]	adj.无错的，无误的	1B
essential	[ɪ'senʃl]	adj.基本的；必不可少的；根本的　　n.必需品；基本知识	3A
estimate	['estɪmət]	n.估计，估算；评价	4A
estimation	[ˌestɪ'meɪʃn]	n.评价，判断；估算	5B
ethic	['eθɪk]	n.道德规范；伦理标准	5B

（续）

单 词	音 标	意 义	单 元
evaluate	[ɪˈvæljueɪt]	v.估计	3B
evaluation	[ɪˌvæljuˈeɪʃn]	n.评估，评价	4A
event	[ɪˈvent]	n.事件	2A
evidence	[ˈevɪdəns]	n.证据 v.证明	6B
exact	[ɪgˈzækt]	adj.精确的；确切的	3B
examine	[ɪgˈzæmɪn]	v.检查，调查	1B
excellent	[ˈeksələnt]	adj.极好的，优秀的	11A
exceptional	[ɪkˈsepʃənl]	adj.例外的	1B
execute	[ˈeksɪkjuːt]	v.执行，实施	10A
execution	[ˌeksɪˈkjuːʃn]	n.实施，执行	2A
exemplify	[ɪgˈzemplɪfaɪ]	v.是……的典范；例证	7B
exhaustive	[ɪgˈzɔːstɪv]	adj.详尽的，彻底的	5B
exhibit	[ɪgˈzɪbɪt]	v.表现；展览	2A
expand	[ɪkˈspænd]	v.扩大，扩展	4A
expect	[ɪkˈspekt]	v.预期；盼望；料想	7A
expedite	[ˈekspədaɪt]	vt.加快进展；迅速完成	11B
experience	[ɪkˈspɪəriəns]	n.经验；经历 v.经历，感受	11A
expert	[ˈekspɜːt]	n.专家，行家 adj.行家的；专业的	5A
explanation	[ˌekspləˈneɪʃn]	n.解释，说明；理由	3A
explanatory	[ɪkˈsplænətri]	adj.解释性的，说明性的	7A
explicit	[ɪkˈsplɪsɪt]	adj.易于理解的；明确的	5B
explorative	[eksˈplɒrətɪv]	adj.探索的	9A
exportable	[eksˈpɔːtəbəl]	adj.可导出的，可输出的	11B
expose	[ɪkˈspəʊz]	n.暴露，揭露	5B
expressivity	[ˌekspreˈsɪvəti]	n.表现力，表达性	2B
extension	[ɪkˈstenʃn]	n.延伸；扩大	3A
extensive	[ɪkˈstensɪv]	adj.广阔的；广泛的；大量的	11A
extra	[ˈekstrə]	adj.额外的	10A
extract	[ˈekstrækt]	v.提取，提炼	5A
factor	[ˈfæktə]	n.因素	1B
failure	[ˈfeɪljə]	n.失败，故障	4A
fascinate	[ˈfæsɪneɪt]	v.深深吸引；迷住	1A
fatigue	[fəˈtiːg]	n.疲劳，厌倦	5A
feature	[ˈfiːtʃə]	n.特征，特点 v.以……为特色	1B
federation	[ˌfedəˈreɪʃn]	n.联邦；同盟，联合会	2B
feed	[fiːd]	v.输送，供应	8B
feedback	[ˈfiːdbæk]	n.反馈	10A

（续）

单 词	音 标	意 义	单 元
feed-forward	[fi:d 'fɔ:wəd]	n.前馈	8A
filter	['fɪltə]	n.过滤器；筛选（过滤）程序 v.过滤	5B
fine-tune	[ˌfaɪn tju:n]	vt.微调	9A
fingerprint	['fɪŋgəprɪnt]	n.指纹，指印	9A
finite	['faɪnaɪt]	adj.有限的；限定的	4B
flag	[flæg]	n.标识，标记 v.标示，标记	6A
flexibility	[ˌfleksə'bɪləti]	n.灵活性；弹性	5B
flexible	['fleksəbl]	adj.灵活的	10B
flourish	['flʌrɪʃ]	vi.繁荣；活跃	1A
fluent	['flu:ənt]	adj.流畅的；流利的	2B
foreseeable	[fɔ:'si:əbl]	adj.可预见到的	1A
form	[fɔ:m]	n.表格；类型；形态，外形	12A
format	['fɔ:mæt]	n.格式 vt.使格式化	5A
forum	['fɔ:rəm]	n.论坛	11A
framework	['freɪmwɜ:k]	n.构架；框架；（体系的）结构	5B
frontier	['frʌntɪə]	n.边界；前沿；新领域	4B
fulfill	[fʊl'fɪl]	vt.履行，执行	8B
function	['fʌŋkʃn]	n.功能；函数	1A
functionality	[ˌfʌŋkʃə'næləti]	n.功能；功能性	11A
fundamental	[ˌfʌndə'mentl]	adj.基础的	1A
fuzzification	[fʌzɪfɪ'keɪʃn]	n.模糊性	3B
fuzzy	['fʌzi]	adj.模糊的	3B
gadget	['gædʒɪt]	n.小装置；小配件	12A
gameplay	['geɪmpleɪ]	n.（计算机游戏的）游戏情节设计，玩法	12A
gap	[gæp]	n.缺口；差距	12A
gather	['gæðə]	v.收集，搜集	9A
generalization	[ˌdʒenrəlaɪ'zeɪʃn]	n.归纳；一般化；普通化	3A
generalize	['dʒenrəlaɪz]	v.概括，归纳	9A
generic	[dʒə'nerɪk]	adj.通用的；一般的	12B
genre	['ʒɒnrə]	n.类型，种类	9A
geographical	[ˌdʒi:ə'græfɪkl]	adj.地理的	2B
geolocalisation	['dʒi:əʊˌləʊkəlaɪ'zeɪʃn]	n.地理定位	5B
giant	['dʒaɪənt]	n.巨人；大公司	11B
glimpse	[glɪmps]	n.一瞥，扫视	1A
goal-oriented	[gəʊl 'ɔ:riəntɪd]	adj.面向目标的，目标导向的	10A
gradually	['grædʒuəli]	adv.逐渐地，逐步地	4B
graph	[grɑ:f]	n.图	4B

（续）

单　词	音　标	意　　义	单　元
ground-breaking	['graʊnd breɪkɪŋ]	adj.开拓性的，独创的	1A
growth	[grəʊθ]	n.发展，增加，增长	7A
guarantee	[ˌgærən'ti:]	v.确保　　n.保证；保修单	2A
guidance	['gaɪdns]	n.指导	11A
habit	['hæbɪt]	n.习惯	10A
hamper	['hæmpə]	vt.妨碍，束缚，限制	1B
handle	['hændl]	v.操作；处理　　n.句柄；手柄	1B
hardware	['hɑ:dweə]	n.硬件	12A
harness	['hɑ:nɪs]	vt.利用；控制	10B
hassle	['hæsl]	n.困难的事情；麻烦的事情	11A
hazard	['hæzəd]	vt.冒险，使遭受危险　　n.危险；冒险的事	3B
heterogeneous	[ˌhetərə'dʒi:nɪəs]	adj.异构的，各种各样的	2B
heuristic	[hjʊ'rɪstɪk]	adj.启发式的	2A
histopathology	[ˌhɪstəʊpə'θɒlədʒi]	n.组织病理学	9A
homonym	['hɒmənɪm]	n.同形同音异义词	9B
homophone	['hɒməfəʊn]	n.同音异义词	9B
humanity	[hjʊ'mænəti]	n.人类；人性；人道	3B
humongous	[hjʊ'mʌŋgəs]	adj.极大的，奇大无比的	5B
hype	[haɪp]	n.炒作，天花乱坠的广告宣传　　vt.大肆宣传；夸张地宣传	8B
hyperplane	['haɪpə,pleɪn]	n.超平面	7A
hypothesis	[haɪ'pɒθəsɪs]	n.假设，假说	3A
hypothetical	[ˌhaɪpə'θetɪkl]	adj.假设的，假定的，假想的	1A
identify	[aɪ'dentɪfaɪ]	vt.识别，认出；确定	5A
idiom	['ɪdɪəm]	n.习语，成语	9B
ignition	[ɪg'nɪʃn]	n.（汽油发动机的）发火装置	3B
imagination	[ɪˌmædʒɪ'neɪʃn]	n.想象，想象力	1A
imbalance	[ɪm'bæləns]	n.不平衡	6A
imitate	['ɪmɪteɪt]	vt.模仿，效仿	1A
implement	['ɪmplɪment]	vt.实施，执行	4B
implication	[ˌɪmplɪ'keɪʃn]	n.含义，蕴涵，蕴含	1A
impossible	[ɪm'pɒsəbl]	adj.不可能的	7A
imprecise	[ˌɪmprɪ'saɪs]	adj.不精确的，不确定的	3B
improve	[ɪm'pru:v]	v.改进，提高	6B
inability	[ˌɪnə'bɪləti]	n.无能，无力	3B
inaccuracy	[ɪn'ækjərəsi]	n.不准确，误差	1B
inadequate	[ɪn'ædɪkwət]	adj.不充足的；不适当的	2A
inappropriate	[ˌɪnə'prəʊpriət]	adj.不恰当的，不适宜的	9B

（续）

单　词	音　标	意　义	单　元
inbuilt	['ɪnbɪlt]	*adj.*嵌入的，内置的	5A
incident	['ɪnsɪdənt]	*n.*事件，事变；敌对行动	6A
incomplete	[,ɪnkəm'pli:t]	*adj.*不完备的；不完全的	3A
incompleteness	[,ɪnkəm'pli:tnəs]	*n.*不完备性，不完整性	4B
inconsistency	[,ɪnkən'sɪstənsi]	*n.*不一致，不协调；前后矛盾	6B
incorporate	[ɪn'kɔ:pəreɪt]	*vt.*包含；使混合；使具体化　　*vi.*包含；吸收；合并；混合	3B
increase	[ɪn'kri:s]	*v.*增加，增长	6B
	['ɪnkri:s]	*n.*增加，增长	
incredibly	[ɪn'kredəbli]	*adv.*难以置信地，很，极为	6B
incur	[ɪn'kɜ:]	*vt.*招致，引起	8B
indicate	['ɪndɪkeɪt]	*v.*表明；指示	6A
indicator	['ɪndɪkeɪtə]	*n.*指示器；指标	12A
inductive	[ɪn'dʌktɪv]	*adj.*归纳的；归纳法的	3A
inefficient	[,ɪnɪ'fɪʃnt]	*adj.*无效率的，无能的	2A
infection	[ɪn'fekʃn]	*n.*传染，感染	5A
infer	[ɪn'fɜ:]	*v.*推断，推理	3A
inference	['ɪnfərəns]	*n.*推理；推断；推论	5A
infinite	['ɪnfɪnət]	*adj.*无限的，无穷的	4B
influence	['ɪnfluəns]	*v.*影响；支配	1B
influx	['ɪnflʌks]	*n.*流入，注入；汇集	8B
infobox	['ɪnfəubɒks]	*n.*信息框	2B
informal	[ɪn'fɔ:ml]	*adj.*非正式的	3A
infrastructure	['ɪnfrəstrʌktʃə]	*n.*基础设施，基础架构	11B
infrequent	[ɪn'fri:kwənt]	*adj.*稀少的；罕见的	7A
ingest	[ɪn'dʒest]	*vt.*接收；吸收；采集；获取	6A
inheritable	[ɪn'herɪtəbl]	*adj.*可继承的，会遗传的	2A
inheritance	[ɪn'herɪtəns]	*n.*继承	2A
initiate	[ɪ'nɪʃieɪt]	*vt.*启动，开始，发起	10A
innovation	[,ɪnə'veɪʃn]	*n.*改革，创新；新观念；新发明	6B
input	['ɪnpʊt]	*n.&v.*输入	5A
insight	['ɪnsaɪt]	*n.*洞察力	11B
inspect	[ɪn'spekt]	*v.*检查；视察	7B
inspire	[ɪn'spaɪə]	*v.*启发；激励	8A
instance	['ɪnstəns]	*n.*实例	2A
instant	['ɪnstənt]	*adj.*即时的，立即的	11B
instantaneously	[,ɪnstən'teɪniəsli]	*adv.*即刻地	12B
instinctual	[ɪn'stɪŋktʃuəl]	*adj.*本能的	7B

（续）

单 词	音 标	意 义	单 元
institution	[ˌɪnstɪˈtjuːʃn]	n.机构	4A
instruct	[ɪnˈstrʌkt]	vt.教导，指导	1A
integrate	[ˈɪntɪɡreɪt]	v.合并；集成	5B
intellectual	[ˌɪntəˈlektʃuəl]	adj.智力的；有才智的 n.知识分子；脑力劳动者	1B
intelligence	[ɪnˈtelɪdʒəns]	n.智能，智力	1A
intensity	[ɪnˈtensəti]	n.强度；烈度	3B
interactive	[ˌɪntərˈæktɪv]	adj.交互式的；互动的	11A
interconnect	[ˌɪntəkəˈnekt]	vi.互相连接，互相联系 vt.使互相连接；使互相联系	1A
interface	[ˈɪntəfeɪs]	n.界面；接口	11B
interoperability	[ˈɪntərˌɒpərəˈbɪləti]	n.互用性，协同工作的能力	2B
interpersonal	[ˌɪntəˈpɜːsnl]	adj.人与人之间的；人际的	12A
interpret	[ɪnˈtɜːprət]	v.解释，诠释；领会	2A
interpretable	[ɪnˈtɜːprətəbl]	adj.能说明的，能解释的，可判断的	7A
interpretation	[ɪnˌtɜːprəˈteɪʃn]	n.解释，说明	2A
interpreter	[ɪnˈtɜːprɪtə]	n.解释器，解释程序	11A
intersect	[ˌɪntəˈsekt]	v.相交，交叉	4B
intervention	[ˌɪntəˈvenʃn]	n.介入，干涉，干预	6A
intonation	[ˌɪntəˈneɪʃn]	n.语调，声调	9B
introduction	[ˌɪntrəˈdʌkʃn]	n.介绍；引进；推出；推行	7A
intrusion	[ɪnˈtruːʒn]	n.入侵，打扰	12A
intuitive	[ɪnˈtjuːɪtɪv]	adj.直观的；直觉的	11A
invalidate	[ɪnˈvælɪdeɪt]	vt.使无效；使作废；证明……错误	3A
invaluable	[ɪnˈvæljuəbl]	adj.非常宝贵的；无价的	11A
invent	[ɪnˈvent]	vt.发明，创造	1A
investigative	[ɪnˈvestɪɡətɪv]	adj.调查性质的；研究的	6B
investment	[ɪnˈvestmənt]	n.投资	7A
irregularity	[ɪˌreɡjəˈlærəti]	n.不规则，无规律	9B
irrelevant	[ɪˈreləvənt]	adj.不相干的；不恰当的	8B
iteration	[ˌɪtəˈreɪʃn]	n.迭代；循环	4A
iteratively	[ˈɪtəˌreitivli]	adv.迭代地	8A
judgemental	[dʒʌdʒˈmentl]	adj.判断的；裁决的	1B
judgment	[ˈdʒʌdʒmənt]	n.判断，鉴定；辨别力，判断力	2A
keyboard	[ˈkiːbɔːd]	n.键盘	10A
knowledge	[ˈnɒlɪdʒ]	n.知识	1A
knowledge-base	[ˈnɒlɪdʒ beɪs]	n.知识库	2A
lag	[læɡ]	vi.走得极慢，落后	11A
layer	[ˈleɪə]	n.层，层次 v.分层	6B

（续）

单　词	音　标	意　义	单　元
layman	['leɪmən]	n.门外汉，外行	1A
lemmatization	[lemətaɪ'zeɪʃn]	n.词形还原	9B
leverage	['li:vərɪdʒ]	n.杠杆作用　　v.利用	12B
library	['laɪbrəri]	n.库	11A
likelihood	['laɪklihʊd]	n.可能性	6B
limitation	[ˌlɪmɪ'teɪʃn]	n.限制，局限；极限	1A
linear	['lɪniə]	adj.线性的	6A
link	[lɪŋk]	v.连接　　n.超文本链接	2B
list	[lɪst]	n.列表；清单	4A
logical	['lɒdʒɪkl]	adj.逻辑（上）的；符合逻辑的	1B
loop	[lu:p]	n.循环	4A
low-cost	[ˌləʊ 'kɒst]	adj.低成本的，价格便宜的	10B
machine-readable	[məʃi:n 'ri:dəbl]	adj.机器可读的，可用计算机处理的	2B
mainstream	['meɪnstri:m]	n.主流；主要倾向，主要趋势	7A
maintainable	[meɪn'teɪnəbl]	adj.可维护的	11A
manual	['mænjuəl]	adj.用手的；手动的，手工的　　n.手册；指南	5A
manufacturer	[ˌmænju'fæktʃərə]	n.制造商，制造厂	1B
map	[mæp]	v.映射	3B
massive	['mæsɪv]	adj.巨大的，大量的	6A
match	[mætʃ]	v.使相配，使相称	3B
materialize	[mə'tɪəriəlaɪz]	vi.具体化；实质化	2B
mathematical	[ˌmæθə'mætɪkl]	adj.数学的；精确的	1B
matrix	['meɪtrɪks]	n.矩阵	5B
meaningful	['mi:nɪŋfl]	adj.意味深长的；有意义的	1A
measure	['meʒə]	v.衡量；测量；量度；估量	4A
measurement	['meʒəmənt]	n.量度，测量	8B
mechanism	['mekənɪzəm]	n.机制	2A
media	['mi:dɪə]	n.媒体	5B
medicine	['medsn]	n.医学；药物	5A
memory	['meməri]	n.存储器，内存	4B
mental	['mentl]	adj.智力的	3A
menu	['menju:]	n.菜单	6B
merger	['mɜ:dʒə]	n.融合，合并	12B
message	[ˌmesɪdʒ]	n.信息；电邮　　v.给……发消息	6A
metadata	['metədeɪtə]	n.元数据	2B
meta-knowledge	['metə 'nɒlɪdʒ]	n.元知识	2A
metaphor	['metəfə]	n.隐喻，暗喻	9B

（续）

单 词	音 标	意 义	单 元
methodology	[ˌmeθəˈdɒlədʒi]	n.方法学；方法论	7B
micro-controller	[ˌmɪkrəʊ kənˈtrəʊlə]	n.微控制器	3B
mimic	[ˈmɪmɪk]	vt.模仿	1A
minimal	[ˈmɪnɪml]	adj.最小的，极小的；极少的	4B
mirror	[ˈmɪrə]	vt.反映；反射	2B
misinterpret	[ˌmɪsɪnˈtɜːprət]	vt.误解，曲解	8B
mislabel	[mɪsˈleɪbl]	v.标记错误	8B
mistake	[mɪˈsteɪk]	n.错误，过失　　v.弄错，误解	1B
moderate	[ˈmɒdərət]	adj.中等的	7A
moderation	[ˌmɒdəˈreɪʃn]	n.适度；自我节制；稳定	5B
modify	[ˈmɒdɪfaɪ]	v.修改，改变	2A
modular	[ˈmɒdjələ]	adj.模块化的	8A
momentum	[məˈmentəm]	n.动量	3B
monitor	[ˈmɒnɪtə]	v. 监视；控制；监测	12A
monotonic	[ˌmɒnəˈtɒnɪk]	adj.单调的，无变化的	3A
mood	[muːd]	n.情绪	12A
moral	[ˈmɒrəl]	n.道德　　adj.道德的	1B
morph	[mɔːf]	vt.改变	12B
multi-dimensional	[ˌmʌltɪ daɪˈmenʃənl]	adj.多维的；多重的	5B
multilateral	[ˌmʌltiˈlætərəl]	adj.多方面的，多边的	2B
navigation	[ˌnævɪˈgeɪʃn]	n.导航	3A
nefarious	[nɪˈfeəriəs]	adj.极坏的，恶毒的	12A
negative	[ˈnegətɪv]	adj.负面的，消极的	1B
network	[ˈnetwɜːk]	n.网络	12B
neural	[ˈnjʊərəl]	adj.神经的	6A
neuron	[ˈnjʊərɒn]	n.神经元；神经细胞	8A
node	[nəʊd]	n.节点	4A
noise	[nɔɪz]	n.噪声；干扰信息	3B
non-deterministic	[nɒn dɪˌtɜːmɪˈnɪstɪk]	adj.非确定性的，不确定的	7A
non-essential	[ˌnɒn ɪˈsenʃl]	adj.不重要的，非本质的	1B
non-linear	[nɒn ˈlɪniə]	adj.非线性的	7A
non-parametric	[ˌnɒn pærəmˈetrɪk]	adj.无参数的	7B
nuance	[ˈnjuːɑːns]	n.细微差别	11A
object	[ˈɒbdʒɪkt]	n.对象；物体；目标	2A
object-oriented	[ˈɒbdʒɪkt ɔːriəntɪd]	adj.面向对象的	11A
obscure	[əbˈskjʊə]	adj.不清楚的；隐蔽的　　vt.使难理解；掩盖；隐藏	5B
observable	[əbˈzɜːvəbl]	adj.可观察的	10A

单　词	音　标	意　义	单　元
observation	[ˌɒbzə'veɪʃn]	n.观察；监视；评论	3A
obstacle	['ɒbstəkl]	n.障碍（物）	8B
occurrence	[ə'kʌrəns]	n.发生，出现	7A
online	[ˌɒn'laɪn]	adj.在线的；联网的；联机的	11A
on-premise	[ɒn 'premɪs]	n.本机端，本地	11B
on-site	['ɒn saɪt]	adj.现场的	12B
ontology	[ɒn'tɒlədʒi]	n.本体论，实体论	2B
operate	['ɒpəreɪt]	v.运转；操作	1B
opinion	[ə'pɪnjən]	n.意见	6A
opponent	[ə'pəʊnənt]	n.对手；敌手	12A
opportunity	[ˌɒpə'tjuːnəti]	n.机会；时机	4A
opposite	['ɒpəzɪt]	n.对立物　adj.对面的；截然相反的	9B
optimal	['ɒptɪməl]	adj.最佳的，最优的	4B
optimality	[ɔpti'mæliti]	n.最优性；最佳性	4B
optimisation	[ɒptɪmaɪ'zeɪʃən]	n.最优法；最优化	5B
optimize	['ɒptɪmaɪz]	vt.使最优化	10A
option	['ɒpʃn]	n.选项，选择；选择权	11B
orchestration	[ˌɔːkɪ'streɪʃn]	n.编排；管弦乐编曲	12B
outcome	['aʊtkʌm]	n.结果	12B
outlay	['aʊtleɪ]	n.花费；费用	12B
outlier	['aʊtˌlaɪə]	n.离群值；异常值	7B
output	['aʊtpʊt]	n.&v.输出	5A
overfit	['əʊvəfɪt]	v.过拟合，过度拟合	6A
overuse	[ˌəʊvə'juːs]	vt.过度使用	9B
panacea	[ˌpænə'siːə]	n.灵丹妙药	3B
paradigm	['pærədaɪm]	n.范式	12A
parallelize	['pærəleˌlaɪz]	vt.使并行，使平行	7A
parameter	[pə'ræmɪtə]	n.参数，变量；限制因素，决定因素	1A
partially	['pɑːʃəli]	adv.部分地	3B
particular	[pə'tɪkjələ]	adj.特定的，特殊的	4A
partition	[pɑː'tɪʃn]	vt.分段，分开，隔开　n.隔离物	7A
pattern	['pætn]	n.模式	9A
pedigree	['pedɪɡriː]	n.血统；家谱　adj.纯种的	6A
perceive	[pə'siːv]	v.察觉，发觉	9A
perception	[pə'sepʃn]	n.洞察力；知觉	2A
perceptron	[pə'septrɒn]	n.感知器	8A
perform	[pə'fɔːm]	v.执行；起……作用	4A

（续）

单 词	音 标	意 义	单 元
performance	[pə'fɔ:məns]	n.性能；执行	2A
personalise	['pɜ:sənəlaɪz]	vt.个性化；个人化	1A
perspective	[pə'spektɪv]	n.观点，看法	1A
phishing	['fɪʃɪŋ]	n.网络钓鱼，网络仿冒	9B
phoneme	['fəʊni:m]	n.音位，音素	9A
pinpoint	['pɪnpɔɪnt]	vt.确定 adj.精确的，精准的；详尽的	8B
pipeline	['paɪplaɪn]	n.管道；渠道	11B
pixel	['pɪksl]	n.像素	5B
plan	[plæn]	n.计划；规划	12B
platform	['plætfɔ:m]	n.平台	5B
poll	['pəʊl]	v.投票；轮询	7B
popularity	[ˌpɒpju'lærəti]	n.流行性，普及度	11A
portable	['pɔ:təbl]	adj.轻便的，可移植的	11A
portal	['pɔ:tl]	n.门户，入口	12A
positive	['pɒzətɪv]	adj.正面的	6B
possess	[pə'zes]	v.拥有；具备	8A
possibility	[ˌpɒsə'bɪləti]	n.可能性；机会；潜力	1A
post	[pəʊst]	n.（论坛等的）帖子	9B
potential	[pə'tenʃl]	n.潜力；可能性 adj.潜在的	10B
practice	['præktɪs]	n.实践	5A
precise	[prɪ'saɪs]	adj.清晰的；精确的	10B
precisely	[prɪ'saɪsli]	adv.精确地	4A
pre-configured	[pri:kən'fɪgəd]	adj.预先配置的	11B
pre-defined	[pri: dɪ'faɪnd]	adj.预定义的	7A
predetermine	[ˌpri:dɪ'tɜ:mɪn]	v.预先决定；事先安排	4B
predictable	[prɪ'dɪktəbl]	adj.可预测的，可预报的	10A
prediction	[prɪ'dɪkʃn]	n.预测，预报	1A
predictor	[prɪ'dɪktə]	n.预言者，预报器	7A
pre-integrated	[pri: 'ɪntɪgreɪtɪd]	adj.预先集成的	11B
premise	['premɪs]	n.前提	3A
prepare	[prɪ'peə]	v.把……准备好，为……做准备	6A
preposition	[ˌprepə'zɪʃn]	n.介词；前置词	3A
preprocess	[pri:'prəʊses]	vt.预处理，预加工	9A
prerequisite	[ˌpri:'rekwəzɪt]	n.先决条件，前提，必要条件	8B
presumption	[prɪ'zʌmpʃn]	n.推测，设想	3A
pre-trained	['pri:treɪnd]	adj.预先训练的	11B
primitive	['prɪmətɪv]	n.原语	9A

（续）

单 词	音 标	意 义	单 元
prior	['praɪə]	adj.优先的；占先的	6A
priority	[praɪ'ɒrəti]	n.优先级，优先权	4A
privacy	['praɪvəsi]	n.隐私	11B
probabilistic	[,prɒbəbə'lɪstɪk]	adj.基于概率的；盖然论的	3A
procedural	[prə'siːdʒərəl]	adj.程序的；过程的	2A
processor	['prəʊsesə]	n.处理器，数据处理机	12A
productivity	[,prɒdʌk'tɪvəti]	n.生产率，生产力	9B
proficient	[prə'fɪʃnt]	adj.精通的，熟练的	3B
profit	['prɒfɪt]	n.利润；好处　v.获益；得益于	10B
program	['prəʊgræm]	n.程序　v.给……编写程序	5A
programmable	['prəʊgræməbl]	adj.可编程的	11B
programmer	['prəʊgræmə]	n.程序设计者，程序员	11A
progress	['prəʊgres]	n.进步；前进；进展	9B
progressively	[prə'gresɪvli]	adv.日益增加地；逐步地	6B
prohibit	[prə'hɪbɪt]	v.禁止，阻止	5B
project	['prɒdʒekt]	n.项目，工程	11A
prominent	['prɒmɪnənt]	adj.突出的，杰出的	11B
promising	['prɒmɪsɪŋ]	adj.有希望的；有前途的	4A
promotion	[prə'məʊʃn]	n.（商品等的）促销，推广	7A
proof	[pruːf]	n.证据；证明；检验	3A
propagation	[,prɒpə'geɪʃn]	n.传播，传输	8A
propel	[prə'pel]	vt.推进；推动	10B
properly	['prɒpəli]	adv.正确地；完全地；真正地	6A
property	['prɒpəti]	n.特性，属性	1A
protect	[prə'tekt]	v.保护	5B
prove	[pruːv]	v.验证，证实	2B
provision	[prə'vɪʒn]	v.为……提供所需物品	12B
proximity	[prɒk'sɪməti]	n.接近，邻近；接近度	6A
psychology	[saɪ'kɒlədʒi]	n.心理学	1A
puzzle	['pʌzl]	n.谜；疑问；智力游戏　v.迷惑；苦苦思索	1A
quantifier	['kwɒntɪfaɪə]	n.量词	2A
query	['kwɪəri]	n.询问；问号　v.查询，询问	1B
questionnaire	[,kwestʃə'neə]	n.调查表；调查问卷	5A
queue	[kjuː]	n.队列	4A
radar	['reɪdɑː]	n.雷达	10A
random	['rændəm]	adj.随机的；任意的	7B
randomize	['rændəmaɪz]	v.使随机化	6A

（续）

单　词	音　标	意　义	单　元
rapid	['ræpɪd]	adj.快速的；瞬间的	7A
rarely	['reəli]	adv.少有地；罕见地	6A
rationally	['ræʃnəli]	adv.讲道理地，理性地	3A
readable	['riːdəbl]	adj.易读的；易懂的，可读的	11A
readjust	[ˌriːə'dʒʌst]	v.再调整	1B
realm	[relm]	n.领域，范围	8B
real-time	[ˌriːl'taɪm]	adj.实时的	9A
reap	[riːp]	v.获得；得到（报偿）	3B
receive	[rɪ'siːv]	v.收到，得到	6B
recognise	['rekəgnaɪz]	vt.辨识，认出，识别出	8A
recognition	[ˌrekəg'nɪʃn]	n.识别	1A
recognize	['rekəgnaɪz]	vt.认出；识别	12A
recommendation	[ˌrekəmen'deɪʃn]	n.推荐；建议	6A
reconstruction	[ˌriːkən'strʌkʃn]	n.重建，再现；重建物，复原物	5B
record	['rekɔːd]	n.记录	6A
recurrent	[rɪ'kʌrənt]	adj.循环的，复现的；周期性的	8A
recursion	[rɪ'kɜːʃn]	n.递归，递归式；递推	4B
recursive	[rɪ'kɜːsɪv]	adj.递归的；回归的	4B
reduce	[rɪ'djuːs]	v.减少，缩小	1B
redundancy	[rɪ'dʌndənsi]	n.冗余，过多	1B
reference	['refrəns]	v.引用　n.提及；查询；征求　adj.供参考的	5B
refine	[rɪ'faɪn]	vt.提炼；改善	6A
reflect	[rɪ'flekt]	v.反映，反射	4A
reflex	['riːfleks]	n.反射作用；反应能力	10A
regression	[rɪ'greʃn]	n.回归	7A
regular	['regjələ]	adj.有规律的；定期的	5A
regularity	[ˌregju'lærəti]	n.规则性，规律性	9A
regularly	['regjələli]	adv.有规律地；经常地	1B
regulation	[ˌregju'leɪʃn]	n.规章，规则　adj.规定的	10B
regulatory	['regjələtəri]	adj.监管的	2B
reinforce	[ˌriːɪn'fɔːs]	vt.强化，增强	6A
reinforcement	[ˌriːɪn'fɔːsmənt]	n.强化，加强，增强	6A
relationship	[rɪ'leɪʃnʃɪp]	n.关系	9A
relevancy	['reləvənsi]	n.关联；恰当；关联事物	7A
relevant	['reləvənt]	adj.有关的，相关联的	5A
reliable	[rɪ'laɪəbl]	adj.可信赖的；可靠的	5A
reliably	[rɪ'laɪəbli]	adv.可靠地，确实地	9B

（续）

单　　词	音　　标	意　　义	单　元
remote	[rɪˈməʊt]	adj.远程的，遥远的	1B
render	[ˈrendə]	v.造成；给予；表达	7A
repeat	[rɪˈpiːt]	v.重复	4B
repeatable	[rɪˈpiːtəbl]	adj.可重复的	11B
repeatedly	[rɪˈpiːtɪdli]	adv.反复地，重复地	6B
repetitive	[rɪˈpetətɪv]	adj.重复的	10A
replace	[rɪˈpleɪs]	v.替换；以……取代；更新	4B
report	[rɪˈpɔːt]	n.报告	12A
repository	[rɪˈpɒzətri]	n.仓库	11A
represent	[ˌreprɪˈzent]	v.代表；相当于；描绘	3A
representation	[ˌreprɪzenˈteɪʃn]	n.表示，表现；陈述	2A
representative	[ˌreprɪˈzentətɪv]	adj.典型的；有代表性的	9A
reproducible	[ˌriːprəˈdjuːsəbl]	adj.可再生的，可复写的；能繁殖的	5A
reputation	[ˌrepjuˈteɪʃn]	n.名誉，名声	11A
rescue	[ˈreskjuː]	v.营救，救助　　n.营救（行动）	7A
researcher	[rɪˈsɜːtʃə]	n.研究者	1A
resistance	[rɪˈzɪstəns]	n.电阻	8A
resolve	[rɪˈzɒlv]	vi.解决	5A
resource	[rɪˈsɔːs]	n.资源	5B
respond	[rɪˈspɒnd]	v.响应；回答，回复	5A
response	[rɪˈspɒns]	n.响应，反应	10A
responsive	[rɪˈspɒnsɪv]	adj.响应的，应答的	5A
responsiveness	[rɪˈspɒnsɪvnəs]	n.响应性	12B
restriction	[rɪˈstrɪkʃn]	n.限制，限定	2A
restructure	[ˌriːˈstrʌktʃə]	v.重构，重建，重组	10B
retrain	[ˌriːˈtreɪn]	vt.重新训练，再教育	12A
retrieval	[rɪˈtriːvl]	n.检索	10A
return	[rɪˈtɜːn]	v.返回	4A
reuse	[ˌriːˈjuːz]	vt.复用，重用	2B
robotics	[rəʊˈbɒtɪks]	n.机器人技术	10B
robust	[rəʊˈbʌst]	adj.健壮的，强健的；结实的	3B
round-the-clock	[ˌraʊnd ðəˈklɒk]	adj.全天候的；不分昼夜的；连续不停的	1B
route	[ruːt]	n.路	4A
routine	[ruːˈtiːn]	n.例程；惯例，常规　　adj.常规的，日常的，平常的	5B
rule	[ruːl]	n.规则　　v.控制，支配	2A
runtime	[ˈrʌntaɪm]	n.运行期，运行时间	2A
sample	[ˈsɑːmpl]	n.样本　　vt.取样	6A

（续）

单 词	音 标	意 义	单 元
sarcasm	['sɑːkæzəm]	*n.*讥讽；嘲讽	9B
satisfy	['sætɪsfaɪ]	*v.*（使）满意，满足	2B
scalable	['skeɪləbl]	*adj.*可扩展的，可升级的	12B
scan	[skæn]	*v.*扫描	5B
scanner	['skænə]	*n.*扫描器；扫描设备	5B
scenario	[sə'nɑːrɪəʊ]	*n.*情景；设想	5A
scheme	[skiːm]	*n.*方案；计划	3A
screen	[skriːn]	*v.*筛选	5B
seamless	['siːmləs]	*adj.*无缝的	12B
search	[sɜːtʃ]	*v.&n.*搜索	4A
select	[sɪ'lekt]	*v.*选择，（在计算机屏幕上）选定	4A
self-aware	[ˌself ə'weə]	*adj.*自知的，自我意识的	1A
semantic	[sɪ'mæntɪk]	*adj.*语义的，语义学的	2A
semi-supervised	['semi ʌn'suːpəvaɪzd]	*adj.*半监督的	6A
sense	[sens]	*n.*感觉官能（即视、听、嗅、味、触五觉）；感觉 *v.*感觉到；意识到	6B
sensor	['sensə]	*n.*传感器	9A
sentence	['sentəns]	*n.*判断	3A
serious	['sɪərɪəs]	*adj.*严重的；令人担忧的	5B
server	['sɜːvə]	*n.*服务器	12B
service	['sɜːvɪs]	*n.*服务	11B
set	[set]	*n.*集合；一套/副/组	1B
share	[ʃeə]	*v.*共有，合用，分享	5B
showcase	['ʃəʊkeɪs]	*v.*展示（优点） *n.*（商店或博物馆等的）玻璃柜台，玻璃陈列柜；展示（优点的）场合	1A
sign	[saɪn]	*n.*迹象；符号	12A
significant	[sɪɡ'nɪfɪkənt]	*adj.*重要的；显著的	11B
similarity	[ˌsɪmə'lærəti]	*n.*相似度	7A
simplicity	[sɪm'plɪsəti]	*n.*简单，朴素	11A
simultaneous	[ˌsɪml'teɪnɪəs]	*adj.*同时的	4B
singleton	['sɪŋɡltən]	*n.*单独	3B
singular	['sɪŋɡjələ]	*adj.*单个的	1A
situation	[ˌsɪtʃu'eɪʃn]	*n.*情况，情景	10A
skill	[skɪl]	*n.*技能，技巧；本领	1B
slot	[slɒt]	*n.*槽	2A
slowdown	['sləʊdaʊn]	*n.*减速，减缓	1B
slur	[slɜː]	*vt.*含糊地说	9B
smartness	['smɑːtnəs]	*n.*聪明，机灵，敏捷	1B

（续）

单 词	音 标	意 义	单 元
smoothly	['smuːðli]	adv.平滑地；流畅地；平稳地	2A
snapshot	['snæpʃɒt]	n.快照	6B
software	['sɒftweə]	n.软件	12A
solution	[sə'luːʃn]	n.解决办法；答案	4B
solve	[sɒlv]	v.解决；破解	1A
sophisticate	[sə'fistɪkeɪtɪd]	adj.有经验的，老于世故的	6B
sophisticated	[sə'fistɪkeɪt]	adj.先进的；复杂的；老练的；见多识广的	12A
sought-after	['sɔːtɑːftə]	adj.很吃香的，广受欢迎的	12A
soundscape	['saʊndskeɪp]	n.音景	12A
spam	[spæm]	n.垃圾邮件	6A
span	[spæn]	v.横跨，跨越	12B
specialist	['speʃəlɪst]	n.专家；专科医生	6B
specification	[ˌspesɪfɪ'keɪʃn]	n.规格；详述；说明书	5A
spectrum	['spektrəm]	n.光谱，波谱；范围；系列	5B
spot	[spɒt]	v.注意到	12A
stable	['steɪbl]	adj.稳定的；牢固的	11A
stack	[stæk]	n.堆栈	11A
stage	[steɪdʒ]	n.阶段	1A
stagger	['stægə]	vt.使吃惊	8B
standalone	['stændə,ləʊn]	adj.单独的，独立的	11A
standard	['stændəd]	n.标准　　adj.标准的	3B
standardization	[ˌstændədaɪ'zeɪʃn]	n.标准化，规格化；规范化	2B
standpoint	['stændpɔɪnt]	n.立场，观点	1A
state	[steɪt]	n.状态	4A
statistical	[stə'tɪstɪkl]	adj.统计的，统计学的	9A
stem	[stem]	n.词干	9B
stimulate	['stɪmjuleɪt]	vt.刺激；激励；促进	1A
stock	[stɒk]	n.股份，股票；库存	6B
store	[stɔː]	v.存储，保存；记忆	4B
strategy	['strætədʒi]	n.策略；部署；战略	4B
streamline	['striːmlaɪn]	vt.流畅；使简单化，使现代化	6B
strengthen	['streŋθn]	v.加强；巩固；支持；壮大	12B
stringent	['strɪndʒənt]	adj.严格的	10B
structural	['strʌktʃərəl]	adj.结构（上）的	2A
sub-agent	[sʌb'eɪdʒənt]	n.子智能体	10A
subclass	['sʌbklɑːs]	n.子类	2B
subgraph	['sʌbgrɑːf]	n.子图	4B

（续）

单　词	音　标	意　　义	单　元
subjective	[səb'dʒektɪv]	*adj.*主观的；个人的	7A
sub-region	[sʌb'riːdʒən]	*n.*子域	2B
subscription	[səb'skrɪpʃn]	*n.*订阅	12B
subset	['sʌbset]	*n.*子集	5B
substitute	['sʌbstɪtjuːt]	*v.*用……代替　*n.*代替者；替代物	8B
subtask	['sʌbtɑːsk]	*n.*子任务	8A
subtly	['sʌtəli]	*adv.*巧妙地	7B
successor	[sək'sesə]	*n.*后继者，继任者	4A
sufficient	[sə'fɪʃnt]	*adj.*足够的，充足的，充分的	11A
suggestion	[sə'dʒestʃən]	*n.*建议；表明	5A
suitably	['suːtəbli]	*adv.*适当地，适宜地；相配地；合适地	2A
summarize	['sʌməraɪz]	*vt.*总结，概述	9B
superclass	['sjuːpəklɑːs]	*n.*超类	2B
superset	['sjuːpəset]	*n.*超集，扩展集，父集	7A
supervise	['suːpəvaɪz]	*v.*监督；管理；指导	6A
supervisor	['suːpəvaɪzə]	*n.*管理者，监督者	8A
supplier	[sə'plaɪə]	*n.*供应商，供应者	10B
surpass	[sə'pɑːs]	*vt.*超过，优于，胜过	1A
surveillance	[sɜː'veɪləns]	*n.*盯梢，监督；监视	5B
susceptible	[sə'septəbl]	*adj.*易受影响的	5B
suspicious	[sə'spɪʃəs]	*adj.*可疑的；不信任的	5A
sustainability	[sə,steɪnə'bɪləti]	*n.*持续性	12B
switch	[swɪtʃ]	*v.*切换；改变，转变	4A
symbol	['sɪmbl]	*n.*符号，记号	2A
symptom	['sɪmptəm]	*n.*症状；征兆	5A
synapse	['saɪnæps]	*n.*（神经元的）突触	8A
synopsis	[sɪ'nɒpsɪs]	*n.*摘要，梗概；大纲	9B
syntactic	[sɪn'tæktɪk]	*adj.*句法的	9A
syntax	['sɪntæks]	*n.*语法	11A
synthesis	['sɪnθəsɪs]	*n.*综合	12B
systematically	[,sɪstə'mætɪkli]	*adv.*有系统地；有组织地	2A
tag	[tæg]	*vt.*加标签于，标注　*n.*标签	6A
tailor	['teɪlə]	*v.*专门制作；定做	6A
Tangle	['tæŋgl]	*n.*纠缠；混乱　*v.*（使）缠结，（使）乱作一团	2B
Task	[tɑːsk]	*n.*工作，任务　*vt.*交给某人（任务）	1A
taxonomy	[tæk'sɒnəmi]	*n.*分类法，分类学，分类系统	2B
telecom	['telɪkɒm]	*n.*电信	10B

（续）

单 词	音 标	意 义	单 元
template	['templeɪt]	n.样板；模板	11B
tempo	['tempəʊ]	n.节奏，拍子	12A
tendency	['tendənsi]	n.倾向，趋势	8B
tension	['tenʃn]	n.焦虑；冲突	5A
terminate	['tɜːmɪneɪt]	v.结束；使终结	4A
text	[tekst]	n.文本	1B
threat	[θret]	n.威胁	5B
threshold	['θreʃhəʊld]	n.阈值	8A
time-consuming	['taɪm kən‚sjuːmɪŋ]	adj.费时的，耗时的	10A
tiredness	['taɪədnəs]	n.疲劳，倦怠	1B
title	['taɪtl]	n.标题　vt.加标题	8B
tone	[təʊn]	n.语调；风格	9B
toolchain	['tuːltʃeɪn]	n.工具链	11B
toolset	['tuːlset]	n.成套工具，工具箱	11B
track	[træk]	v.跟踪，追踪	8B
trait	[treɪt]	n.特点，特征，特性	9A
transform	[træns'fɔːm]	v.转化	11B
transitive	['trænzətɪv]	adj.传递的	2B
translate	[trænz'leɪt]	v.（使）转变为	5B
transparent	[træns'pærənt]	adj.透明的	10B
trapezoidal	[træpɪ'zɔɪdəl]	adj.梯形的	3B
traverse	[trə'vɜːs]	v.遍历	4A
treatment	['triːtmənt]	n.治疗	10A
tremendous	[trə'mendəs]	adj.极大的，巨大的	10B
trial	['traɪəl]	v.测试　n.试验	6A
triangular	[traɪ'æŋgjələ]	adj.三角形的	3B
tricky	['trɪki]	adj.难办的，棘手的	11A
triple	['trɪpl]	n.三元组	2B
triplestore	['trɪplstɔː]	n.三元存储	2B
trustworthy	['trʌstwɜːði]	adj.值得信赖的，可靠的	5B
tuning	['tjuːnɪŋ]	n.调整	7A
unauthorized	[ʌn'ɔːθəraɪzd]	adj.未经授权的，未经许可的	12B
unavailable	[‚ʌnə'veɪləbl]	adj.难以获得的，不能利用的	8B
unaware	[‚ʌnə'weə]	adj.不知道的；未察觉到的；不注意的	6B
uncertain	[ʌn'sɜːtn]	adj.不确定的；多变的	3A
uncertainty	[ʌn'sɜːtnti]	n.不确定；不可靠	5A
uncover	[ʌn'kʌvə]	v.揭示，发现	9A

（续）

单 词	音 标	意 义	单 元
undeniable	[ˌʌndɪ'naɪəbl]	adj.不可否认的，无可争辩的	12B
understandable	[ˌʌndə'stændəbl]	adj.能懂的，可理解的	5A
undertake	[ˌʌndə'teɪk]	v.承担；从事	11B
undetected	[ˌʌndɪ'tektɪd]	adj.未被察觉的，未发现的	6B
unemployment	[ˌʌnɪm'plɔɪmənt]	n.失业；失业率	1B
unfamiliar	[ˌʌnfə'mɪliə]	adj.不熟悉的；不常见的	9A
unguided	['ʌn'gaɪdɪd]	n.无向导的；不能控制的	4A
unhindered	[ʌn'hɪndəd]	adj.不受妨碍的，不受阻碍的	12B
unification	[juːnɪfɪ'keɪʃn]	n.统一，联合；一致	2B
unlabeled	[ʌn'leɪbld]	adj.未标记的	6A
unnecessary	[ʌn'nesəsəri]	adj.不需要的；没必要的	1B
unplanned	[ˌʌn'plænd]	adj.无计划的，未筹划的	10B
unpredictable	[ˌʌnprɪ'dɪktəbl]	adj.不可预测的	1B
unrestricted	[ˌʌnrɪ'strɪktɪd]	adj.不受限制的；无限制的；自由的	12B
unsafe	[ʌn'seɪf]	adj.不安全的，危险的	12B
unsettlingly	[ʌn'setlɪŋli]	adv. 令人不安地	12A
unsupervise	[ʌn'suːpəvaɪz]	v.无监督；不管理	6A
untapped	[ʌn'tæpt]	adj.未开发的，未利用的	8B
update	[ʌp'deɪt]	vt.更新，升级	12A
upload	[ʌp'ləʊd]	vt.上传，上载	5B
usefulness	['juːsflnəs]	n.有用，有益，有效	8B
utility	[juː'tɪləti]	n.效用，功用	10A
vacuum	['vækjuːm]	n.真空；清洁 v.用真空吸尘器清扫	6A
vague	[veɪg]	adj.模糊的；（思想上）不清楚的；（表达或感知）含糊的	3B
valid	['vælɪd]	adj.有效的；正当的	3A
valuable	['væljuəbl]	adj.有价值的	9A
value	['væljuː]	n.价值	8B
variation	[ˌveəri'eɪʃn]	n.变体；变化，变动	11B
variety	[və'raɪəti]	n.多样，多样化	8B
vector	['vektə]	n.向量，矢量	8A
velocity	[və'lɒsəti]	n.高速	8B
veracity	[və'ræsəti]	n.真实	8B
versatile	['vɜːsətaɪl]	adj.多用途的；多功能的	7B
versatility	[ˌvɜːsə'tɪləti]	n.多用途性	5B
vertex	['vɜːteks]	n.顶点	4B
video	['vɪdiəʊ]	n.视频	5B
videogame	['vɪdiəʊgeɪm]	n.电子游戏	12A

（续）

单 词	音 标	意 义	单 元
virtual	['vɜ:tʃuəl]	adj.（计算机）虚拟的；实质上的，事实上的	1A
visible	['vɪzəbl]	adj.可见的	6B
vision	['vɪʒn]	n.视觉，视力	5B
visual	['vɪʒuəl]	adj.视觉的	5B
visualization	[ˌvɪʒuəlaɪ'zeɪʃn]	n.可视化	11A
voice	[vɔɪs]	n.语音	6B
volume	['vɒlju:m]	n.大量	8B
web	[web]	n.网络	1A
weight	[weɪt]	n.权重	6A
willingness	['wɪlɪŋnəs]	n.愿意，乐意	5B
wireless	['waɪələs]	adj.无线的	12A
workflow	['wɜ:kfləʊ]	n.工作流程	11A
workload	['wɜ:kləʊd]	n.工作量，工作负担	12B
workplace	['wɜ:kpleɪs]	n.工作场所，车间；工厂	12A
worst-case	['wɜ:st keɪs]	adj.最坏情况的	4B

附录 B 词 组 表

词 组	意 义	单 元
a group of	一群，一组	5A
a range of	一系列；一些；一套	5B
a series of	一系列；一连串	8A
a set of	一套，一组	1A
a variety of	各种各样的	6B
abductive reasoning	溯因推理	3A
abstract notion	抽象概念	9A
according to	根据	4B
act on	按照……而行动；遵行	5B
actual value	实际值	8A
adapt to	适应	1B
adaptable for	适合于	5B
adaptive resonance theory network	自适应共振理论网络	8A
adverse drug reaction	药物不良反应	7A
affective computing	情感计算	1A
AI-based model	基于人工智能的模型	1B
AI-driven device	由人工智能驱动的设备	10A
altitude control	高度控制	3B
an array of	一排，一群，一批	4A

（续）

词　　组	意　　义	单　元
analyze text	分析文本	11B
architecture diagram	体系结构图	3B
artificial intelligence	人工智能（AI）	1A
artificial neuron	人工神经元	8A
asset management	资产管理	10B
association algorithm	关联算法	6A
association rule	关联规则	6A
attached to	附属于	1B
augmented reality	增强现实	12A
automated reasoning	自动推理	2A
autonomous driving	自动驾驶	10A
back propagation	反向传播	6B
backward chaining	后向链接	5A
backward search	反向搜索	4B
base on	基于	5A
batch processing	批处理	12B
Bayes probability theorem	贝叶斯概率定理	7A
be applied to	适用于，应用于	3B
be based on	基于	1A
be compatible with	与……兼容	5B
be defined as	被定义为	3B
be described as	被描述为	10A
be designed for	为……而设计	5A
be divided into	划分为	3A
be equivalent to	等于，等同于	4B
be explained as	被解释为	3A
be filled with	充满着	9B
be fit into	适合；融入	7B
be obtained by	由……得到，由……获得	4B
be referred to as	被称作，被称为	1A
be saved into	被保存到	4B
be similar to	类似于，相似于	4B
be stuck in	困于，停止不前，动弹不得	4B
be suitable for	适用于	11A
best first search	最佳优先搜索	4A
bidirectional search	双向搜索	4B
big data	大数据	1A
binary variable	二元变量	7B
black comedy	黑色喜剧	9A

（续）

词 组	意 义	单 元
blind search	盲目搜索	4B
blog post	博文，博客帖子	6A
bond rating	债券评级，债券分级	7B
bottom-up reasoning	自底向上推理	3A
braking system	制动系统	3B
branching factor	分支因子	4B
break down	分解	8A
brute force	暴力，蛮力	4B
build model	构建模型	11A
bump sensor	碰撞传感器	10A
business acumen	商业头脑，业务头脑	8B
business logic	业务逻辑	11A
business process	业务流程，业务过程	12A
business-critical data	关键业务数据	12B
call out	调来；召集	6A
carry out	执行；进行；完成	1B
cause-effect reasoning	因果关系推理	3A
chronic disease	慢性病	7B
classification tree	分类树	7B
cloud computing	云计算	11B
cloud platform	云平台	11B
cloud provider	云技术提供商	12B
cloud search	云搜索	11B
cloud storage	云存储	11B
clustered system	集群系统	11A
clustering algorithm	聚类算法	6A
cognitive computing	认知计算	12B
come up with	想出，想到	11A
common element	共同元素	9A
common interface	通用接口	2B
common sense reasoning	常识推理	3A
computation power	计算能力	11B
computational resource	计算资源	1A
computer science	计算机科学	1A
computer-readable language	计算机可读语言	5B
computing resource	计算资源	12B
concern with	与……有关；关心……	9B
concurrency support	并发支持	11A
condition monitoring	状态监测	10B

（续）

词 组	意 义	单 元
consist of	包含；由……组成	5A
consumer good	消费品	10B
container orchestration	容器编排	12B
continuous variable	连续变量	7B
contracts analysis	合约分析	11B
contribute to	贡献；有助于	5B
convert ... to	把……转换为……	11B
convert into	转换为	9A
co-reference resolution	指代消歧，共指解析	9B
cost function	成本函数	8A
credit card	信用卡	6A
credit scoring	信用评分，资信评分	7B
credit scoring system	信用评分制度	7B
current location	当前位置	4A
current node	当前节点	4B
current state	当前状态	4A
customer service	客户服务	1B
customer support	客户支持	10A
customized response	客户化响应，自定义响应	6B
cyber security	网络安全	10B
dark area	盲点，暗区	10B
data center	数据中心	12B
data communication system	数据通信系统	10B
data exploration	数据探索	7B
data management paradigm	数据管理范式	2B
data mining	数据挖掘	6A
data object	数据对象	10A
data schema	数据模式	2B
data science	数据科学	4A
data serialization	数据序列化	2B
data snapshot	数据快照	6B
data structure	数据结构	4B
data visualization	数据可视化	11A
database management system	数据库管理系统	4A
data-driven algorithm	由数据驱动的算法	12A
day-to-day task	日常任务	1B
deal with	处理	3A
debated topic	争论的话题	12A
decision rule	决策规则	6A

（续）

词 组	意 义	单 元
decision tree	决策树	6A
declarative knowledge	陈述性知识	2A
deductive reasoning	演绎推理	3A
deep AI	深度人工智能	1A
deep feed-forward neural network	深度前馈神经网络	8A
deep learning	深度学习	1A
deep neural network	深度神经网络	6A
dense depth estimation	密集深度估计	5B
depend on	根据；依据，依靠	1A
dependent variable	因变量，应变数，因变数	6A
derived computed data	派生出的计算数据	8B
design process	设计过程	10B
deterministic inference engine	确定性推理引擎	5A
diagnose malfunction	诊断故障	7B
digital assistance	数字助理	1B
digital assistant	数字助理，数字助手	6A
digital era	数字时代	8B
digital image	数字图像	5B
digital twin	数字孪生	10B
dirt detection sensor	污垢检测传感器	10A
driverless car	无人驾驶汽车	6B
due to	由于，因为	1B
economic benefit	经济效益，经济利益	12B
end up	最终处于，到头来	8B
end user	最终用户	5A
equal variance	等方差，同方差	7B
error value	误差值	8A
extraction service	提取服务	11B
face detector	人脸检测器	5B
face recognition	面部识别，人脸识别	9A
facial recognition	面部识别	1A
factual knowledge	事实知识	5A
feature extraction	特征提取	9A
feature hierarchy	特征层次	8A
feature space	特征空间	7B
feature vector	特征向量	7A
fill in	填写；填补	12A
financial institution	金融机构	6B
find out	发现，找出	4B

（续）

词　　组	意　　义	单　元
fingerprint identification	指纹识别	9A
fingerprint lock	指纹锁	9A
flick through	浏览	12A
flooded with	淹没；挤满，充满	5B
focus on	聚焦	5B
formal logic	形式逻辑	2A
formal semantic	形式语义	2B
forward chaining	前向链接	5A
forward propagation	前向传播	6B
forward search	正向搜索	4B
frame representation	框架表示	2A
fraud detection	欺诈检测	6A
fuel injection quantity	燃油喷射量	3B
fuzzy logic system	模糊逻辑系统	3B
fuzzy set	模糊集合	3B
general-graph search	通用图搜索	4B
general-purpose language	通用语言	11A
generic rule	通用规则	3A
globally unique identifier	全局唯一标识符	2B
globular cluster	球形聚类	7A
goal node	目标节点	4A
goal state	目标状态	4A
goal-based agent	基于目标的智能体	10A
gold mine	金矿	8B
grammatical tagging	语法标注	9B
graph representation	图形表示	2B
graphical network	图形化的网络	2A
greedy best-first search	贪婪最佳优先搜索	4A
greedy search	贪婪搜索	4A
hand in hand	手拉手，携手；密切合作	11A
handwriting recognition	手写识别	7B
have impact on	对……产生冲击；碰撞，影响	1B
heterogeneous data	异构数据	2B
heuristic function	启发函数	4A
heuristic knowledge	启发式知识	2A
heuristic rule	启发式规则	3A
heuristic value	启发值	4A
hidden layer	隐藏层	6A
hidden pattern	隐藏模式，隐含模式	8B

（续）

词　组	意　义	单　元
hierarchical clustering	层次聚类	7A
higher-level agent	高级智能体	10A
high-performance scientific computing	高性能科学计算	11A
high-powered processor	高性能处理器	12A
historical data	历史数据	12B
household appliance	家用电器	12A
human resource	人力资源	1B
hybrid cloud	混合云	12B
hypothesis testing	假设检验	1A
image recognition	图像识别	8B
independent variable	自变量，自变数	6A
inductive reasoning	归纳推理	3A
industrial robot	工业机器人	10B
industry standard	行业标准	12B
inference engine	推理工具，推理机	3B
inferential knowledge	推理知识	2A
infinite loop	无限循环，无穷循环	4B
information-heavy service	信息密集型服务	2B
informed search	知情搜索	4A
inheritable knowledge	可继承的知识	2A
initial phase	初起阶段	5B
initial state	初始状态，初态	4A
input layer	输入层	6A
instance relation	实例关系	2A
instance-based algorithm	基于实例的算法	6A
instant access	即时访问	11B
intelligent agent	智能体；智能代理	2A
interact with	与……相互作用，与……相互影响，与……相互配合	5A
interconnected node	互联的节点	6B
intermediate value	中间值	3B
internet of things	物联网	5B
investment portfolios	投资组合	6B
item set	项目集	7A
iterative method	迭代法	7A
iterative process	迭代过程	6A
judgemental power	判断能力	1B
keep in mind	牢记	8B
K-means clustering	K 均值聚类	7A
knowledge acquisition	知识获得，知识收集	5A

词　　组	意　　义	单　元
knowledge domain	知识领域	2B
knowledge mining	知识挖掘	11B
knowledge representation	知识表示	1A
Kohonen self-organizing maps	科霍宁自组织映射	8A
labeled data	已标记的数据	5B
large-scale problem	大规模问题，大型问题	4A
lazy learner	惰性学习器	7B
leaf node	叶节点	7B
linear algebra	线性代数	5B
linear regression	线性回归	6A
linear relationship	线性关系	7B
load capacity	负载容量；载重量	3B
loan approval	贷款批准	6B
logical reasoning	逻辑推理	2A
logistic regression	逻辑回归	6A
lower-level agent	低级智能体	10A
low-power device	低功耗设备	12A
low-skilled job	低技能工作	1B
machine aided surgery	机器辅助手术	5B
machine intelligent	机器智能	1A
machine learning	机器学习（ML）	1A
machine learning algorithm	机器学习算法	5B
machine translation	机器翻译	9B
main processing unit	主处理单元，主处理机	5A
market demand	市场需求	1B
master data	主数据	2B
membership function	隶属函数	3B
memory cell	记忆细胞，记忆单元	8A
memory power	存储能力	1A
memory requirement	内存需求（量）	4A
missing value	缺失值	7B
mission-critical business process	执行关键任务的业务流程	9B
mixed initiative system	混合主动系统	8B
model-based reflex agent	基于模型的反射型智能体	10A
modeling instrument	建模工具	2B
monotonic reasoning	单调推理	3A
multi-dimensional data	多维数据	5B
multilayer perceptron	多层感知器	8A
multi-layered neural network	多层神经网络	6B

（续）

词　　组	意　　义	单　元
multi-nominal logistic regression	多项逻辑回归	7B
naive Bayes classifier algorithm	朴素贝叶斯分类器算法	7A
narrow down	变窄，减少，缩小	4A
natural language	自然语言	2A
nefarious activity	恶意活动	12A
negative emotion	消极情绪，负面情绪	7A
network data structure	网络数据结构	2B
neural network	神经网络	6A
neural pattern recognition	神经模式识别	9A
news publishing industry	新闻出版行业	2B
non-decreasing function	非递减函数	4B
non-monotonic reasoning	非单调推理	3A
normal distribution	正态分布	7B
numerical computing	数值计算	11A
object occlusion	目标遮挡	5B
object recognition	物体识别	11B
object-oriented design	面向对象设计	2B
online storage	在线存储	8B
open source	开源	11A
operating cost	运营成本	10A
operating system	操作系统	12B
option pricing	期权定价	7B
ordinal logistic regression	有序逻辑回归	7B
output layer	输出层	6A
panorama construction	全景建筑	5B
path cost	路径成本	4A
pattern recognition	模式识别	5B
period of time	时段，一段时间	5A
personal favourite	个人喜好	5B
personalized data	个性化数据	12B
personalized experience	个性化的体验	12A
platform independence	平台独立性	11A
point out	指出	11A
policy maker	决策者，制定政策者	8B
predicate logic	谓词逻辑	2A
predicted value	预测值	8A
predictive maintenance	预测性维护	10B
predictive modeling	预测建模	8B
preventive measures	防护性措施，预防措施	7B

（续）

词　　组	意　　义	单　元
pre-written code	预先编写的代码	11A
private cloud	私有云，专用云	12B
private network	私有网络，专用网络	12B
probabilistic inference engine	概率推理引擎	5A
procedural knowledge	程序性知识	2A
product review	产品评论	8B
production line	生产线，流水线	12A
production rule	产生式规则	2A
programming language	程序设计语言，编程语言	2A
programming tool	程序设计工具，编程工具	7A
propositional logic	命题逻辑	3A
public cloud	公共云	12B
pure heuristic search	纯启发式搜索	4A
put ... into	把……译成……	9B
QR code	二维码	5B
radial basis function	径向基函数	8A
random forest	随机森林	7B
random sample	随机样本	7B
raw data	原始数据	12B
raw input	原始输入	3B
reach up to	高达	1B
real time	实时	6B
recognize-act cycle	识别-行动循环	2A
recurrent neural network	循环神经网络	8A
regression algorithm	回归算法	6A
regression tree	回归树	7B
reinforcement machine learning	强化机器学习	6A
relational method	关系法	2A
relevance rate	相关率	7A
rely on	依靠，依赖	5B
remote sensing	遥感	7B
risk management	风险管理	7B
road sign	交通标志，路标	10A
roll out	推出；铺开	12A
root node	根节点	4B
rule-based modeling	基于规则建模	9B
science fiction novel	科幻小说	1A
search box	搜索框，搜索栏	5A
search engine	搜索引擎	7A

（续）

词　　组	意　　义	单　　元
search space	搜索空间	4A
search tree	搜索树	4A
self-driving car	自动驾驶汽车	1A
semantic modeling	语义建模	11B
semantic network	语义网	2A
semantic reasoning	语义推理	9B
semantic tag	语义标记	2B
semi-supervised machine learning	半监督机器学习	6A
semi-supervised model	半监督模型	12A
sensitive data	敏感数据	12A
sentence parsing	句子解析	9B
sentiment analysis	情绪分析，情感分析	7A
serve as	充当，担任	2B
serverless computing	无服务器计算	12B
set out	列出；安排；摆放；陈列	2A
set theory	集合论	3B
shared understanding	共识	2B
significant role	重要作用	10B
single layer perceptron	单层感知器	8A
smart car manufacturer	智能汽车制造厂	8B
smart factory	智能工厂	10B
smart manufacturing	智能制造	10B
smart playlist	智能播放列表	12A
social intelligence	社会智能	1A
social media	社交媒体	5B
software package	软件包，程序包	11A
software-based robot	基于软件的机器人	12A
space complexity	空间复杂度	4B
spam detection	垃圾邮件检测	9B
spam mail	垃圾邮件	5B
special case	特例	2B
speech recognition	语音识别	1A
split ... into	把……划分为……	9A
squared difference	平方差值	8A
stock market	股票市场，股票行情	7A
store in	存储于，存储在	5A
strong AI	强人工智能	1A
structural knowledge	结构化知识	2A
structural pattern recognition	结构化模式识别	9A

（续）

词 组	意 义	单 元
structured information	结构化信息	9B
structured query	结构化查询	2B
successor node	子节点，后继节点	4B
supervised machine learning	监督机器学习	1A
support vector machine	支持向量机	6A
symmetric relationship	对称关系	2B
technology giant	科技巨头	11B
technology stack	技术栈	11A
testing set	测试集	9A
text analysis technology	文本分析技术	2B
text mining technique	文本挖掘技术	2B
third-party service provider	第三方服务提供商	12B
time complexity	时间复杂度	4B
top-down reasoning	自顶向下推理	3A
traditional technique	传统技术	8B
traffic light	红绿灯，交通灯	5B
training data set	训练数据集	6A
training rule	训练规则	9A
training set	训练集	9A
training subset	训练子集	6A
turn to	向……求助	11A
unauthorized access	未授权的访问	12B
underlying rule	潜在规则	8A
uninformed search	不知情搜索	4B
unsupervised machine learning	无监督机器学习	1A
user experience	用户体验	1A
user interface	用户界面	5A
utility-based agent	基于效用的智能体	10A
vacuum extractor	真空吸尘器	10A
validation set	验证集	9A
venture capital	风险投资，风险资本	10B
virtual assistant	虚拟助手	1A
virtual data layer	虚拟数据层	2B
virtual model	虚拟模型	10B
visible layer	可见层	6B
visual data	可视化数据	5B
vital role	重要作用，重要角色	7A
weak AI	弱人工智能	1A
web page	网页	5B

（续）

词　组	意　义	单　元
web scraping	网络抓取	8B
weighted tree	加权树	4B
wind up	（使自己）陷入，卷入，落得	8B
wireless communications technology	无线通信技术	12A
with respect to	关于，至于，谈到	10B
word segmentation	词切分	9B
work together	协同工作	8B
worst case scenario	最坏情况	4A

附录C　缩　写　表

缩　写	意　义	单　元
AGI (Artificial General Intelligence)	通用人工智能	1A
ANI (Artificial Narrow Intelligence)	窄人工智能	1A
ANN (Artificial Neural Network)	人工神经网络	8A
API (Application Programming Interface)	应用程序接口	11B
AR (Augmented Reality)	增强现实	5B
ASI (Artificial Super-Intelligence)	超人工智能	1A
BFS (Breadth First Search)	宽度优先搜索	4A
CNN (Convolutional Neural Network)	卷积神经网络	6A
CPU (Central Processing Unit)	中央处理器	12B
CV (Computer Vision)	计算机视觉	5B
DBSCAN (Density-Based Spatial Clustering of Applications with Noise)	具有噪声的基于密度的聚类方法	9A
DFS (Depth First Search)	深度优先搜索	4A
DLS (Depth Limited Search)	限制深度搜索	4B
ES (Expert System)	专家系统	5A
FIFO (First Input First Output)	先进先出	4B
GDP (Gross Domestic Product)	国内生产总值	2B
GPS (Global Position System)	全球定位系统	6A
GPU (Graphics Processing Unit)	图形处理单元	11A
IA (Intelligent Agent)	智能体	10A
IaaS (Infrastructure as a Service)	基础设施即服务	12B
IDDFS (Iterative Deepening Depth First Search)	迭代加深深度优先搜索	4B
IIoT (Industrial Internet of Thing)	工业物联网	10B
IoT (Internet of Things)	物联网	1A
IT (Information Technology)	信息技术	3B
JVM (Java Virtual Machine)	Java 虚拟机	11A

缩　写	意　义	单　元
KB (Knowledge Base)	知识库	5A
KG (Knowledge Graph)	知识图谱	2B
KNN (K-Nearest Neighbor)	K 最近邻算法	6A
KRR (Knowledge Representation and Reasoning)	知识表示和推理	2A
LSTM (Long Short Term Memory)	长短期记忆	8A
MRI (Magnetic Resonance Imaging)	磁共振成像	5B
NER (Named Entity Recognition)	命名实体识别	9B
NLG (Natural Language Generation)	自然语言生成	9B
NLP (Natural Language Processing)	自然语言处理	1A
NLTK (Natural Language Toolkit)	自然语言工具包	9B
NLU (Natural Language Understanding)	自然语言理解	11B
NN （Neural Network）	神经网络	8A
ONNX (Open Neural Network Exchange)	开放神经网络交换	11B
OWL (Web Ontology Language)	网络本体语言	2B
PaaS （Platform as a Service）	平台即服务	12B
RDF (Resource Description Framework)	资源描述框架	2B
RNN (Recurrent Neural Network)	递归神经网络	6A
RPM (Revolutions Per Minute)	转数/分	3B
SaaS (Software as a Service)	软件即服务	12B
SDK (Software Development Kit)	软件开发工具包	11B
SVM (Support Vector Machine)	支持向量机	7A
UCS (Uniform Cost Search)	统一成本搜索	4A
VM (Virtual Machine)	虚拟机	11B
VR (Virtual Reality)	虚拟现实	5B
W3C(World Wide Web Consortium)	万维网联盟，又称 W3C 理事会	2B